MOTHER NATURE'S REVENGE

A Theory on Universal Continuity and Sustainability

DR. RONALD BARNES

MOTHER NATURE'S REVENGE
A THEORY ON UNIVERSAL CONTINUITY AND SUSTAINABILITY

Bennett books may be ordered through booksellers or by contacting:

Bennett Media and Marketing
1603 Capitol Ave., Suite 310 A233
Cheyenne, WY 82001
www.thebennettmediaandmarketing.com
Phone: 1-307-202-9292

ISBN: 978-1-957114-71-2 (Paperback)
ISBN: 978-1-957114-72-9 (Hardcover)
ISBN: 978-1-957114-70-5 (eBook)

Printed in the United States of America

Preface

This book is dedicated to Swedish environmentalist Greta Tintin Eleonora Ernman Thunberg in appreciation for her persistent campaign and devotion to the issue of global climate change.

The book is also dedicated to Al Gore, environmental activist and former Vice President of the United States of America. Both individuals have exerted relentless and dedicated effort to create public awareness of climate change and a dedication to solve the problem. Their efforts are a major influence in bringing climate change into the mainstream of American and world concerns as an important issue. President Donald Trump has mocked and criticized the young environmental activist, making the following statements:

> *"What an actress!"*
>
> *'"She's getting the best education socialism can steal. I won't be held hostage by someone who just got a learner's permit."*
>
> *"I don't share exultation about Greta Thunberg." Russian President, Vladimir Putin added: "It's deplorable when someone is using children and teenagers in their interests."*

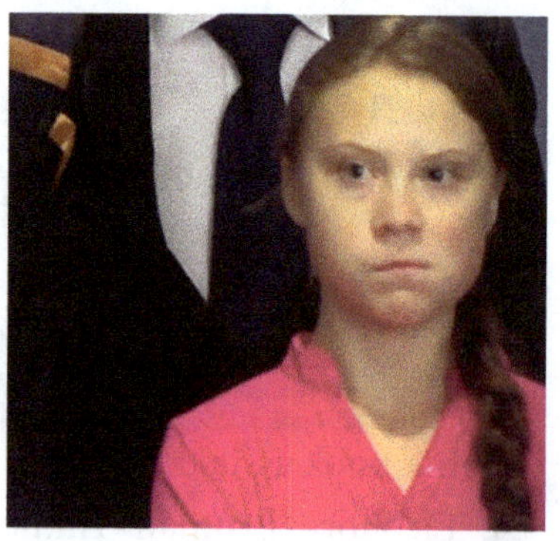

Greta is pictured showing an expression of contempt when Donald Trump passes in her vision. Trump seems to solicit this type of response from many Americans, as well as people around the world. Based on their comments, it is no wonder that Ms. Thunberg has a look of disgust at the sight of these men who ignore the reality of climate change and global warming. The shame of it all is that these grown men, world leaders, downplayed the important message of Thunberg and Gore. What is deplorable is that some of our world leaders are ignorant and mindless about addressing the global warming problem. When they criticize teenagers who herald the cause of global warming because of concern for their future that is a sign of leadership failure. It is deplorable that world leaders show little concern over the future of the Children of the world, and consider themselves leaders. Thunberg is obviously more committed to the world environmental phenomena and the future of the human race than the leaders of the most powerful countries in the world: Trump (past U.S. president and Russian

President Putin. The fact that they give little attention to the future of our world, other than to maintain their own wealth and power, dictate, and build destructive weapons, is a sign that either global warming will devastate the environment, or the world leaders will destroy the world themselves with the use of nuclear weapons. There is an aphorism called Say's Law, a central component to Keynesian economics, associated with the supply-side of economics, which states: "every supply creates its own demand." (Keys, 1936). The existence of nuclear weapons creates the realistic potential to use nuclear weapons.

Miss Greta Thunberg, twice nominated for a Nobel Peace Prize, is gaining worldwide sentiment and respect because she truly cares about the future of our world, obviously more than Trump and Putin, who ironically have also been nominees but never won.

In 2007, former United States Vice-President Al Gore and the IPCC were jointly awarded the Nobel Peace Prize for their efforts to build up and disseminate greater knowledge about the man-made climate change conditions plaguing the earth, and to lay the foundations for the measures that are needed to counteract such change. In March 2010 two nonprofit organizations founded by Al Gore, *The Alliance for Climate Protection* and *The Climate Project*, joined together, and in July 2011 the combined organization was renamed the Climate Reality Project. In February 2012, the Climate Reality Project organized an expedition to the Antarctic with civic and business leaders, activists and concerned citizens from many countries.

The efforts of Thunberg and Gore are especially significant because they represent a generational synergy and ideology alignment between generations. Their efforts contributed to the awareness leading to worldwide collaboration to mitigate environmental pollution. The future generations are in jeopardy of suffering from global warming / climate change and environmental pollution that will deteriorate the quality of their lives. It is noteworthy to recognize the alignment of these two generations on this issue of climate change.

CONTRIBUTIONS

Portions of this book were taken from the Intergovernmental Panel on Climate Change (IPCC) Reports. They have diligently worked on providing information for the international community to address the issue of climate change before catastrophe becomes a reality. The contributors to the IPCC are listed as follows:

Drafting Authors: Hans-O. Pörtner (Germany), Debra C. Roberts (South Africa), Helen Adams (United Kingdom), Carolina Adler (Switzerland/Chile/Australia), Paulina Aldunce (Chile), Elham Ali (Egypt), Rawshan Ara Begum (Malaysia/Australia/Bangladesh), Richard Betts (United Kingdom), Rachel Bezner Kerr (Canada/USA), Robbert Biesbroek (The Netherlands), Joern Birkmann (Germany), Kathryn Bowen (Australia), Edwin Castellanos (Guatemala), Gueladio Cissé (Mauritania/Switzerland/France), Andrew Constable (Australia), Wolfgang Cramer (France), David Dodman (Jamaica/United Kingdom), Siri H. Eriksen (Norway), Andreas Fischlin (Switzerland), Matthias Garschagen (Germany), Bruce Glavovic (New Zealand/South Africa), Elisabeth Gilmore (US A/Canada), Marjolijn Haasnoot (The Netherlands), Sherilee Harper (Canada), Toshihiro Hasegawa (Japan), Bronwyn Hayward (New Zealand), Yukiko Hirabayashi (Japan), Mark Howden (Australia), Kanungwe Kalaba (Zambia), Wolfgang Kiessling (Germany), Rodel Lasco (Philippines), Judy Lawrence (New Zealand), Maria Fernanda Lemos (Brazil), Robert Lempert (USA), Debora Ley (Mexico/Guatemala), Tabea Lissner (Germany), Salvador Lluch-Cota (Mexico), Sina Loeschke (Germany), Simone Lucatello (Mexico), Yong Luo (China),

Brendan Mackey (Australia), Shobha Maharaj (Germany/Trinidad and Tobago), Carlos Mendez (Venezuela), Katja Mintenbeck (Germany), Vincent Möller (Germany), Mariana Moncassim Vale (Brazil), Mike D Morecroft (United Kingdom), Aditi Mukherji (India), Michelle Mycoo (Trinidad and Tobago), Tero Mustonen (Finland), Johanna Nalau (Australia/Finland), Andrew Okem (SouthAfrica/Nigeria), Jean Pierre Ometto (Brazil), Camille Parmesan (France/USA/United Kingdom), Mark Pelling (United Kingdom), Patricia Pinho (Brazil), Elvira Poloczanska (United Kingdom/Australia), Marie-Fanny Racault (United Kingdom/France), Diana Reckien (The Netherlands/Germany), Joy Pereira (Malaysia), Aromar Revi (India), Steven Rose (USA), Roberto SanchezRodriguez (Mexico), E. Lisa F. Schipper (Sweden/United Kingdom), Daniela Schmidt (United Kingdom/Germany), David Schoeman (Australia), Rajib Shaw (Japan), Chandni Singh (India), William Solecki (USA), Lindsay Stringer (United Kingdom), Adelle Thomas (Bahamas), Edmond Totin (Benin), Christopher Trisos (South Africa), Maarten van Aalst (The Netherlands), David Viner (United Kingdom), Morgan Wairiu (Solomon Islands), Rachel Warren (United Kingdom), Pius Yanda (Tanzania), Zelina Zaiton Ibrahim (Malaysia)

Drafting Contributing Authors: Rita Adrian (Germany), Marlies Craig (South Africa), Frode Degvold (Norway), Kristie L. Ebi (USA), Katja Frieler (Germany), Ali Jamshed (Germany/Pakistan), Joanna McMillan (German/Australia), Reinhard Mechler (Austria), Mark New (South Africa), Nick Simpson (South Africa/Zimbabwe), Nicola Stevens (South Africa)

Visual Conception and Information Design: Andrés Alegría (Germany/Honduras), Stefanie Langsdorf (Germany)

Date: 27 February 2022 06:00 UTC

AUTHOR'S PURPOSE

The author's purpose in writing this book is to address and give an additional voice to the critical issue of climate change / global warming confronting our world. People like Al Gore, former Vice President of the United States of America, and Greta Thunberg are addressing global warming on the national and international stage; as is the intergovernmental Panel on Climate Change, a group of experts from countries around the world. However, the details and critical nature of climate change is often conveyed in technical terms and concepts that common people do not grasp or understand. Ordinary people can have a significant impact in mitigating the effects of climate change. It is the intent of this author and this book to describe and convey the concept of climate change / global warming in non-technical terms and concepts the common person can understand. To mitigate the potential destructive impact of climate change / global warming will require behavior change. Behavior change is a function of transformation from old behavior to new behavior. It is also a function of opening the mental state of an individual to think in terms different from one's previous mindset. The process of behavior change requires alignment between the "new" way an individual thinks and the "new" way an individual behaves. Reconstructing human thinking and behavior is a challenging task; however, it can only begin when the individual has information that gives them incentive to change. This intent of this author and this book is to provide the common individual, information in terms they can understand to begin the process of behavior change. Before it is too late.

TABLE OF CONTENTS

Mimi Ayres,
Editor

Introduction

The Theory on Universal Continuity and Sustainability

This book is only one of many contributions that identify the reality and seriousness of climate change in our world. Many science experts, an overwhelming number of experts, and organizations, (Appendix 1) have raised the issue of the immediate necessity to address climate change but to no significant notice on the part of conservative politicians, the rich and wealthy and the companies that pollute our environment. For the average citizen, climate change explained using scientific rationale and descriptions, contains concepts that are difficult to grasp for the average citizen. It is the intent of this book to explain and address the problem of climate change in ways the average, non-scientific person can understand. Greed is a factor inhibiting climate change. It weighs in the failure of American and world politicians to act and support the cause to reverse the warming of our planet. What this book does is correlate the phenomena of climate change to industrialization, population increase, proliferation of fossil fuel emissions, and changes in the environmental phenomena of our world, the increase of natural disasters occurring over the world. The evidence presented in this book reinforces hypothesis there is no coincidence in the relationship between industrialization, population increase, fossil fuel emissions, natural disasters, human behavior, and global warming. There is a causal, stimulus / response relationship between human behavior and global warming / climate change. Why humans fail to address the issue of climate change / global warming in a significantly actionable manner indicates the shortsightedness and incompetence of our world leaders. It also indicates the frivolous casual way some people deal with the issue.

One can hypothesize the value system of our world leaders and several individuals is seriously flawed in preference of money over life. One can also argue that the psychology of politicians is likewise flawed, as they operate with an incompetence that has potential to destroy our world. This book does not claim that global warming is not a significant issue with many individuals. What it does claim is that those individuals (world leaders) are lax in their responsibility to deal with the problem in an urgent manner. Even those world leaders who recognize the problem do not address it with urgency, relative to the crisis it imposes. If someone discovers they have a serious case of cancer, would they put off treatment because they have a vacation planned, or they have a big project at work to complete, or because of what reason? Maybe it depends on how advanced the cancer is. According to experts, the global warming / climate change problem is serious and is, at this moment, affecting the world with detrimental consequences. If an individual's cancer was as serious as the experts indicate climate change / global warming is, then immediate change is necessary.

One of the ways the phenomena of climate change can be explained is by simple observation. Glaciers are melting. Sea levels are rising. Landscapes are deteriorating. Natural disasters are occurring with more frequency (clean water shortages, earthquakes, hurricanes, tornadoes, floods, and wildfires). Experts predict that sea levels are rising and cities are sinking. Unless serious action is taken, experts predict that a number of the world's cities will be underwater by 2100. Meaning they will no longer be inhabitable. Some of the world cities in jeopardy of being impacted by rising sea levels, frequently flooding and being devoured by the sea are: (Nash, 2022).

City	**Population**
Tokyo Japan	37,435,191
Mumbai India	20,185,064
New York City, New York USA	20,140,470

Osaka Japan	19,222,665
Istanbul Turkey	15,415,197
Kolkata India	14,974,073
Jakarta, Indonesia	10,562,088
London, United Kingdom	9,425,622
Amsterdam Netherlands	1,157,519
Cancun Mexico	971,798
Lagos, Nigeria	23,437,435
Houston, Texas USA	2,304,580
Boston Massachusetts USA	4,941,632
Lisbon Portugal	2,971,587
Dubai, UAE	2,921,376
Vancouver Canada	2,606,351
ABU Dhabi UAE	1,511,768
Dhaka, Bangladesh	8,906,039
Ho Chi Minh Vietnam	8,837,544
San Francisco, California USA	7,680,175
Venice, Italy	853,761
Charleston VA USA	799,636
Macau China	661,838
Male Maldives	554,542
Long Beach California USA	466,742
Savanna Georgia USA	404,798
Nassau The Bahamas	385,637
Key West Flordia USA	73,090
Punta Cana Dominican Rep	100,023
Cockburn TN. Turks & Caicos	42,953
Virginia Beach, Virginia USA	1,479,000
Bangkok, Thailand	10,722,815
New Orleans, Louisiana USA	1,270,530
Dublin Ireland	1,241,953
Honolulu Hawaii USA	1,160,508

Rotterdam, Netherlands	651,446
Alexandria, Egypt	5,381,785
Sydney Australia	4,991,654
Miami, Florida USA	6,138,333
Copenhagen Denmark	1,358,608

However, observation to accurately explain phenomenon must be based on logic that makes sense and supported by empirical data. The way psychologist and scientist study, research and collect data on phenomenon is by examining prior studies on a topic, examining data collected by other professionals that is proven reliable and trustworthy, individual interviews, group interviews, questionnaires, by observation, collecting the quantitative data, and monitoring over a period of time. When research data is collected, the psychologist or scientist analyzes the data and draws conclusions that should be reliable and trustworthy based on the way the psychologist or scientist frames, presents and discusses the conclusions. Conclusive findings should be consistent with the information and data collected. That is what the author of this book will achieve in this proposed theory.

The Theory being proposed is:

There is an order to the universe. A natural order that is consistent with every form of organic matter, an order that is perpetuating and sustaining to the earth humans inhabit. When humans ignore the universal order, when their actions are inconsistent with the universal order. The earth will respond. Consequently, the universal order, when violated, responds to the rules of stimulus / response. In the case of earth, the response is for preservation, not for human life but preservation of the earth, itself.

Many of the natural environmental occurrences in our world are consistent with the law of stimulus / response and consistent with universal survival and sustenance. The current environmental occurrence of natural disasters is a result of response to the stimulus of human behaviors, which has influenced the environmental circumstance beyond its natural occurrence. The earth will sustain itself even at the cost of eliminating all life on the planet and regenerating a subsequent species of life. The earth has done it before, and it will do it again for its own continuity and sustainability.

On a simple scale, consider the human body, as an organic mechanism, when cut, infected with virus or disease; has in many cases, an uncanny ability to heal itself, unless the injury is too severe. The fact that humans in leadership have ignored the relational connections in the universe and the world only indicates their shortsighted ignorance.

Edward Thorndike developed **Stimulus response (S-R) theory**. It is associated with the **law of effect, the law of exercise and the law of readiness.** S-R theory states that occurrences (responses) are a reaction to the stimulus (stimulus cause response). This theory emphasizes that positive actions (stimulus) create positive results (response) and negative actions stimulate and influence negative results (response). Thorndike's theory presents the original S-R concept of behavioral psychology: "*Learning is the result of associations formed between stimuli and responses. Such associations or "habits" become strengthened or weakened by the nature and frequency of the S-R pairings*" (Culatta, 2020). Therefore, the results from actions (response) is a factor that determines whether an individual took proper and positive actions (Nazir, 2018). Another of the earliest supporters of S-R theory was psychologist, B. F. Skinner. Thorndike and Skinner both regarded stimulus response as a concept of the learning environment, as being integral to the concept of learning among kids. Positive stimulus is a "*reinforcement when the child realizes the communicative value of words and phrases. Stimulus-response (S-R) theories are central to the principles of learning and conditioning. They are based on the assumption that human behavior is learned*" (Nazir, 2018, p. 153). In plain terms, S-R means that for every stimulus (action), there is a response (re-action) or for every action (stimulus), there is a reaction (response). Responding to positive stimulus in ways that produce positive results can be conditioning that standardizes positive behavior. Continuous response to stimulus that produces negative results indicates a failure to embrace learning. When individuals fail to embrace or produce positive response to stimulus, and acts against the principles of positive stimulus with positive response, continuously, that is an indication of human deficiency and failure to learn. Some call it ignorant, stupid, educationally challenged, or handicapped.

Another perspective on S-R theory is that it is "a theory that proposes all learning consists primarily of the strengthening of the relationship between the stimulus and the response. In developing this theory, Thorndike proposed three laws: the law of effect, the law of exercise, and the law of readiness (Oxford Reference 2020, p. 1). **The law of effect** principle developed by Edward Thorndike suggests that:

> *"Responses that produce a satisfying effect in a particular situation become more likely to occur again in that situation, and responses that produce a discomforting effect become less likely to occur again in that situation (Gray, 2011, p. 108–109)."*

One can argue that humans acting contrary to the law of effect, who ignore the negative discomforting effects of their actions (stimulus), when it comes to the natural environment conditions, they cause negative environmental response. This is contrary to the law of effect that states, "*Responses that produce a discomforting effect become less likely to occur again in that situation*" (Gray, 2011). Human actions affecting the environment produce a negative effect and humans continue to perpetuate the same behaviors. Empirical evidence provided in this document correlates with this theory.

A saying attributed to Albert Einstein that defines insanity is: doing the same thing over and over again in the same way and expecting a different result when it is obvious the intended results are not achieved (Wilczek, 2013).

> *"That one learns by doing and one cannot learn a skill based on behavioral consequence. For instance, by watching others, it is necessary to practice the skill, because by doing so the bond between stimulus and response is strengthened"* (Oxford Reference, 2020).

The law of exercise states:

Humans are continuously giving stimulus consistent with behaviors negatively affecting the environment is an indication; they are not learning or unwilling to learning.

The law of readiness states:

> *"That learning is dependent upon the learner's readiness to act, which facilitates the strengthening of the bond between stimulus and response."*

A noteworthy observation is that failure to take positive steps in addressing the environmental global warming / climate change issue indicates the learner does not appear to possess a readiness to act. Meaning that learning does not take place or response to stimulus is ignored and the bond between Stimulus-Response is not established. In an environmental context, humans continued behavior, ignoring environmental response (natural disasters), will cause continued disasters and suffering.

The alignment of S-R theory to global warming is an indication that S-R has broader application beyond the learning education environment but has a significant relevance to human's ability to learn on a behavioral level. Edward Thorndike developed S-R theory in the context of human interactions with humans and for the educational environment. This author is proposing the argument that S-R theory has application beyond the learning environment. S-R theory has universal application and applies to how organic matter in our universe responds to stimulus and to the proliferation of actions occurring thereafter.

Actions or human behaviors (stimulus) causes obvious responses (reactions) and effects upon the environment and the earth on which we live. Humans acting contrary to the law of effect, defy *Thorndike's concept of connectionism*, and defy the law of exercise. This keeps the law of readiness from engaging, which inhibits learning and allows environmental conditions to get worse. When human stimulus regarding the earth environment is not positively reinforcing, the earth environment acts in accordance and responds to human stimulus in a manner consistent with the theory of Stimulus-Response. The earth responds with natural disasters to human's pollutant stimulus. This author will present evidence and argue that the earth responds to human stimulus in a discomforting effect manifested by earthquakes hurricanes, tornadoes, floods, forest fires and natural disasters. Regardless of the evidence, humans fail to acknowledge the theory of Stimulus-Response when it comes to the environment. The earth responds negative to human behaviors (stimulus) and because of human behavior, the earth responds with discomforting effect, natural disasters. Because humans have ignored the negative effects of their actions (stimulus), they are acting in contradiction to S-R theory, law of effect that states, "responses that produce a discomforting effect become more likely not to occur and re-occur again in that situation" (Gray, 2011). Humans continue to pollute the earth regardless of earth responding disastrously negative. One can argue that is a reason natural disasters are proliferating and increasing in occurrence compared to prior years. Natural disasters are also occurring in off seasons that is unusual according to past occurrences.

Edward Thorndike also developed the philosophy of connectionism which states that learning is a product resulting between stimulus and productive responsive to such stimulus. Stimulus, causing a reaction (response), the response is the reaction to the stimulus. The connection between the stimulus and response (S-R) is called the S-R bond or the S-R connection (Instructional Design.org, n.d.; Study.com, 2020). The bond is either a positive or a negative bond depending on the response to the stimulus. In the case of the environment, the bond established between the human stimulus (human actions) and the response by the environment, is a negative bond, a negative connection because the environmental response to fossil fuel emissions are natural disasters. Simply put, Humans reap what they sow.

The prognosis of human's negative stimulus is an indication that their learning is not taking place. Because humans are not acting in accordance with the theory, of Stimulus-Response, that produces positive environmental response and as a result, environmental conditions are getting worse. Despite warnings and feedback from science experts (Appendix 1) and because conservative authorities; fossil fuel polluters, business people, greedy people and the administration of Donald Trump, who set the time-clock to address climate change / global warming back years. Failure to adhere to expert warnings, cause disasters to happen. As a result, our world is in peril, in the midst of serious and immediate danger of harming or destroying much of the species living on earth. The question to ask is: when the environment becomes too toxic for normal human habitation, what will be done? What good will saying, "I'm sorry" make? I hope that the next mutations or generation of humans will have better sense, better intelligence and concern for more than just **money**.

Einstein did not believe the phenomena in the world was unpredictable, nor did he believe in the inherent unpredictability of the world. . He believed that: "Insanity is doing the same thing over and over and expecting different results" (Wilczek, 2013). Einstein did not believe in the inherent unpredictability of the world, indicating that the natural order of the universe is not like throwing dice (Wilczek, 2013). The world and nature operates in a logical manner consistent with the manner humans interact with the world and the environment.

CHAPTER I

Environmental Terminology

Adaptation Adjustment in natural or human systems to a new or changing environment that exploits beneficial opportunities or moderates negative effects.

Anxiety - (adj. anxious) A nervous or almost fearful reaction to events causing excessive uneasiness and apprehension. People with anxiety may even develop panic attacks.

Asthma - A disease affecting the body's airways, which are the tubes through which animals breathe. Asthma obstructs these airways through swelling, the production of too much mucus or a tightening of the tubes. As a result, the body can expand to breathe in air, but loses the ability to exhale appropriately. The most common cause of asthma is an allergy. It is a leading cause of hospitalization and the top chronic disease responsible for kids missing school.

Atmosphere - The envelope of gases surrounding Earth or another planet.

Behavior - The way a person or other organism acts towards others or conducts itself.

carbon dioxide (or **CO_2**) - A colorless, odorless gas produced by all animals when the oxygen they inhale reacts with the carbon-rich foods that they've eaten. Carbon dioxide also is released when organic matter (including fossil fuels like oil or gas) is burned. Carbon dioxide acts as a greenhouse gas, trapping heat in Earth's atmosphere. Plants convert carbon dioxide into oxygen during photosynthesis, the process they use to make their own food.

Cell - The smallest structural and functional unit of an organism. Typically, too small to see with the naked eye, it consists of watery fluid surrounded by a membrane or wall. Animals are made of anywhere from thousands to trillions of cells, depending on their size. Some organisms, such as yeasts, molds, bacteria and some algae, are composed of only one cell.

Centers for Disease Control and Prevention (or CDC) - An agency of the U.S. Department of Health and Human Services, CDC is charged with protecting public health and safety by working to control and prevent disease, injury and disabilities. It does this by investigating disease outbreaks, tracking exposures by Americans to infections and toxic chemicals, and regularly surveying diet and other habits among a representative cross-section of all Americans.

Chemical - A substance formed from two or more atoms that unite (become bonded together) in a fixed proportion and structure. For example, water is a chemical made of two hydrogen atoms bonded to one oxygen atom. Its chemical symbol is H_2O. Chemical can also be an adjective that describes properties of materials that are the result of various reactions between different compounds.

Chikungunya - A tropical disease that has been crippling large numbers of people in Africa and Asia. It's caused by a virus that is spread by mosquitoes. It recently has been spreading widely throughout warm nations. More than 3 million people have suffered through its initial flu-like symptoms. A large share may also go on to develop intense pain in their muscles and joints that can last months to years. There is no cure or vaccine.

Climate - The weather conditions prevailing in an area in general or over a long period.

Climate Change - Long-term, significant change in the climate of Earth. It can happen naturally or in response to human activities, including the burning of fossil fuels and clearing of forests.

Colleague - Someone who works with another, a co-worker or team member.

Combustion - (adj. combustible) The process of burning.

Compound - (often used as a synonym for chemical) A compound is a substance formed from two or more chemical elements united in fixed proportions. For example, water is a compound made of two hydrogen atoms bonded to one oxygen atom. Its chemical symbol is H_2O.

Control - A part of an experiment where there is no change from normal conditions. The control is essential to scientific experiments. It shows that any new effect is likely due only to the part of the test that a researcher has altered. For example, if scientists were testing different types of fertilizer in a garden, they would want one section of it to remain unfertilized, as the control. Its area would show how plants in this garden grow under normal conditions. And that give scientists something against which they can compare their experimental data.

Development - (in biology) The changes an organism undergoes from conception through adulthood. Those changes often involve chemistry, size and sometimes even shape.

Disorder - (in medicine) A condition where the body does not work appropriately, leading to what might be viewed as an illness. This term can sometimes be used interchangeably with disease.

Disproportionately Exposed or Vulnerable Community A community in which climate change, pollution, or environmental destruction have exacerbated systemic racial, regional, social, environmental, and economic injustices by disproportionately affecting Indigenous peoples, communities of color, migrant communities, deindustrialized communities, depopulated rural communities, the poor, low-income workers, women, the elderly, the unhoused, people with disabilities, or youth.

Drought - An extended period of abnormally low rainfall; a shortage of water resulting from this.

Economics - (adj. **economic**) The social science that deals with the production, distribution and consumption of goods and services and with the theory and management of economies or economic systems. A person who studies economics is an economist.

Environmental health - A research field that focuses on measuring the effects of pollutants and other factors in the environment on the health of people, wildlife or ecosystems.

Environmental Justice (EJ) The fair treatment and meaningful involvement of all people regardless of race, color, culture, national origin, or income, with respect to the development, implementation, and enforcement of environmental laws, regulations, and policies to ensure that each person enjoys (1) the same degree of protection from environmental and health hazards; and (2) equal access to any federal agency action on environmental justice issues in order to have a healthy environment in which to live, learn, work, and recreate.

Environmental Justice Community A community with significant representation of communities of color, low-income communities, or tribal and Indigenous communities, that experiences or is at risk of experiencing higher or more adverse human health or environmental effects.

Equation - In mathematics, the statement that two quantities are equal. In geometry, equations are often used to determine the shape of a curve or surface.

Exhaust - (in engineering) The gases and fine particles emitted — often at high speed and/or pressure — by combustion (burning) or by the heating of air. Exhaust gases are usually a form of waste.

Factor - Something that plays a role in a particular condition or event, a contributor.

Fossil fuel - Any fuel — such as coal, petroleum (crude oil) or natural gas — that has developed in the Earth over millions of years from the decayed remains of bacteria, plants or animals.

Frontline Community A low-income community, community of color, or tribal community that is already or could be disproportionately affected or burdened by climate change and its impacts.

Global warming - The gradual increase in the overall temperature of Earth's atmosphere due to the greenhouse effect. This effect is caused by increased levels of carbon dioxide, chlorofluorocarbons and other gases in the air, many of them released by human activity.

Greenhouse - A light-filled structure, often with windows serving as walls and ceiling materials, in which plants are grown. It provides a controlled environment in which set amounts of water, humidity and nutrients can be applied — and pests can be prevented entry.

Hydrocarbons - Any of a range of large molecules created by chemically bound carbon and hydrogen atoms. Crude oil, for example, is a naturally occurring mix of many hydrocarbons.

Immune system -The collection of cells and their responses that help the body fight off infections and deal with foreign substances that may provoke allergies.

Impair - (n. impairment) To damage or weaken in some way.

Infection - A disease that can spread from one organism to another. It's usually caused by some sort of germ.

Infectious - An adjective that describes a type of germ that can be transmitted to people, animals or other living things.

IPCC - The UN Intergovernmental Panel on Climate Change (IPCC)

The Intergovernmental Panel on Climate Change (IPCC) was set up by the World Meteorological Organization (WMO) and United Nations Environment to provide an objective source of scientific information. In 2013 the IPCC provided more clarity about the role of human activities in climate change when it released its Fifth Assessment Report. It is categorical in its conclusion: climate change is real and human activities are the main cause.

IQ - (or intelligence quotient) A number representing a person's reasoning ability. It's determined by dividing a person's score on a special test by his or her age, then multiplying by 100.

Journal - (in science) A publication in which scientists share their research findings with the public. Some journals publish papers from all fields of science, technology, engineering and math, while others are specific to a single subject. The best journals are peer-reviewed: They send out all submitted articles to outside experts to be read and critiqued. The goal, here, is to prevent the publication of mistakes, fraud or sloppy work.

Link - A connection between two people or things.

Malaria - A disease caused by a parasite that invades the red blood cells. The parasite is transmitted by mosquitoes, largely in tropical and subtropical regions.

Mitigation Measures to reduce the amount and speed of future climate change by reducing emissions of heat trapping gases or removing carbon dioxide from the atmosphere.

Organ - (in biology) Various parts of an organism that perform one or more particular functions. For instance, an ovary is an organ that makes eggs, the brain is an organ that interprets nerve signals and a plant's roots are organs that take in nutrients and moisture.

Ozone - A colorless gas that forms high in the atmosphere and at ground level. When it forms at Earth's surface, ozone is a pollutant that irritates eyes and lungs. It is also a major ingredient of smog.

Particle - A minute amount of something.

Pollutant - A substance that taints something — such as the air, water, our bodies or products. Some pollutants are chemicals, such as pesticides. Others may be radiation, including excess heat or light. Even weeds and other invasive species can be considered a type of biological pollution.

Population - (in biology) A group of individuals from the same species that lives in the same area.

Power Plant - An industrial facility for generating electricity.

Premature - Too early; before something should occur. Premature births, for instance, are when babies are born weeks or months early — potentially before they are ready for life on their own, outside their mom's protective womb.

Range - The full extent or distribution of something. For instance, a plant or animal's range is the area over which it naturally exists. (in math or for measurements) The extent to which variation in values is possible. Also, the distance within which something can be reached or perceived.

Resilience The capability to anticipate, prepare for, respond to, and recover from significant multi-hazard threats with minimum damage to social well-being, health, the economy, and the environment.

Review paper (in science publishing) - A paper that reviews the data and findings in a broad body of work by many research teams. This may include 50 to 200 different research studies or more. The authors then synthesize the findings, looking for patterns that may emerge from the data. These patterns may strengthen — or weaken — the conclusions that seem reasonable when considering just a single paper or two.

Risk - The chance or mathematical likelihood that some bad thing might happen. For instance, exposure to radiation poses a risk of cancer. Or the hazard — or peril — itself. Among cancer risks that the people faced were radiation and drinking water tainted with arsenic.

Society - An integrated group of people or animals that generally cooperate and support one another for the greater good of them all.

Stress - (in biology) A factor, such as unusual temperatures, moisture or pollution, that affects the health of a species or ecosystem. (in psychology) A mental, physical, emotional, or behavioral reaction to an event or circumstance, or stressor, that disturbs a person or animal's usual state of being or places increased demands on a person or animal; psychological stress can be either positive or negative.

Toxic - Poisonous or able to harm or kill cells, tissues or whole organisms. The measure of risk posed by such a poison is its toxicity.

Trillion - A number representing a million million — or 1,000,000,000,000 — of something.

Urban - Of or related to cities, especially densely populated ones or regions where lots of traffic and industrial activity occurs. The development or buildup of urban areas is a phenomenon known as urbanization.

Weather - Conditions in the atmosphere at a localized place and a particular time. It is usually described in terms of particular features, such as air pressure, humidity, moisture, any precipitation (rain, snow or ice), temperature and wind speed. Weather constitutes the actual conditions that occur at any time and place. It's different from climate, which is a description of the conditions that tend to occur in some general region during a particular month or season.

World Health Organization (WHO) - An agency of the United Nations, established in 1948, to promote health and to control communicable diseases. It is based in Geneva, Switzerland. The United Nations relies on the WHO for providing international leadership on global health matters. This organization also helps shape the research agenda for health issues and sets standards for pollutants and other things that could pose a risk to health. "WHO" also regularly reviews data to set policies for maintaining health and a healthy environment.

Chapter II

What people need to know about climate change?

Why Should Humanity be concern about Global warming?

The world we inherited

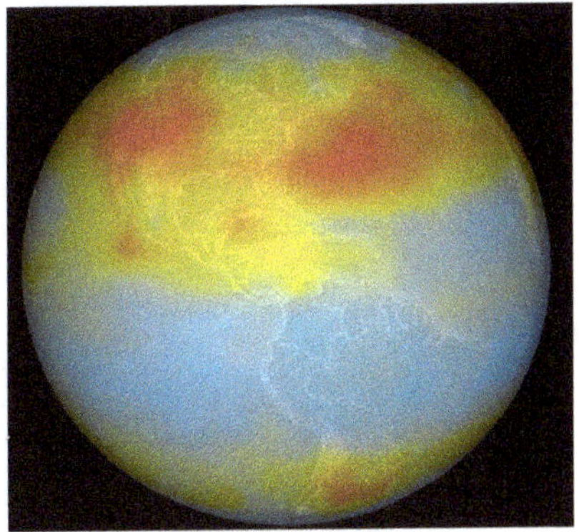

The world we are creating

A Basic Description of the Climate Change Problem

The reality is that global warming and climate Change are real problems. In December 2021, Tornadoes devastated the Midwest United States. In the middle of winter, December 2021, temperatures were 50-60 degrees Fahrenheit. This is extremely rare for this time of the year in this geography. Tornadoes in the Midwest United States winter is a rarity.

Exploring and correlating the global population increase with the increase in industrialization, the increase in fossil fuels polluting the earth, clearly explains the global warming crisis and the rise of natural disasters. Questions to ask are: what is the intelligence quotient of political leaders that they ignore the obvious? Our world leadership is grossly inadequate, not to address this problem with serious conviction. What are their real objectives that they are willing to risk life on earth? Chapter VII of this book discusses the "Denial of climate Change." The chart below indicates the global temperature change since 1880 (NASA, 2021).

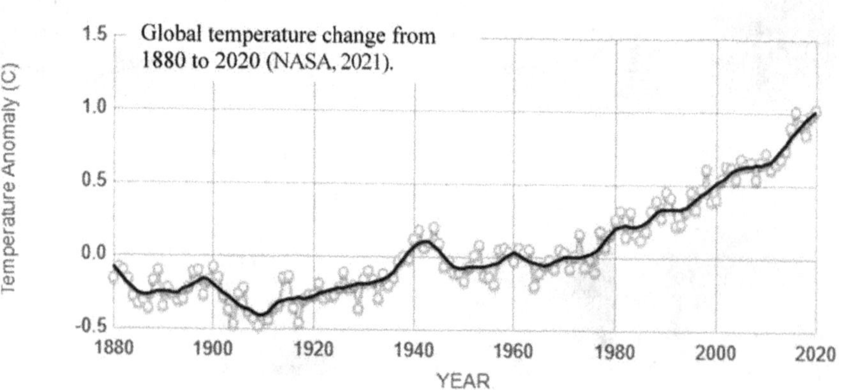

Climate change refers to the fact that, over several years, the average temperature of the earth is changing (climate change). The earth is getting warmer (global warming). The earth warms naturally from the heat of the sun and because warming emissions from natural ocean water and vegetation emerge into the atmosphere giving off heat. Historically these emissions have escaped into the atmosphere with some warming retention on earth necessary to maintain life. The burning of fossil fuels such as coal, oil and gas and the resulting emissions forms an ozone layer, referred to as the greenhouse gas (GHG) effect. When ozone layer prevents heat from escaping into the atmosphere, this is the greenhouse gas (GHG) effect. The GHG's are carbon dioxide, water vapor, methane forming the ozone layer. The greater the buildup of greenhouse gases, the warmer Earth gets. The GHG effect prevents the natural emissions and fossil fuel emissions from being released into the atmosphere. Increased heat is retained within earth's atmosphere and further warms the earth's surface. Fossil fuel emissions are a source of creating the ozone layer which prevents heat from escaping into the atmosphere. Much of the gases from fossil fuels in the atmosphere cause air pollution which can affect breathing. The burning of gas, oil and coal causes a carbon build up in earth's hemispheres. The warming of the earth's surface happens because the ozone layer prevents gases from escaping into the atmosphere. Global warming causes the environment to respond with hurricanes, tornadoes, wildfires, floods, and melting glaciers causing sea levels to rise, warming oceanic conditions causing hurricanes, warming air causing tornadoes, air pollution potentially hazardous to humans and living creatures. Scientists predict that if the average temperature of the earth increases more than 1.5 degrees Celsius by the end of the century, human survival will be in jeopardy.

The Paris Agreement and Climate Change Initiatives (Appendix II)

On December 12, 2015, in Paris, the United Nations Framework Convention on Climate Change came to a landmark agreement. Governments worldwide agreed to the landmark Paris climate accord, setting a goal of keeping global warming well below 2 degrees Celsius with an aim of no higher than 1.5°C. The international community is now struggling to set policies that can achieve those goals. Signed by 196 nations, **the Paris Agreement** (Appendix II) is the first comprehensive global treaty to combat climate change. It will enter into force once it is ratified by at least 55 of the countries, contributing to at least 55% of the global greenhouse gas emissions (Inside Climate News, 2017).

Climate change is defined as gradual changes in all the interconnected weather elements on our planet over approximately 30 years. The data shows the Earth is warming and it's up to us (humans) to make the changes necessary for a healthier planet. The reasons and conditions causing the international community to come together to address the issue of climate change are stated below. Descriptions are taken from references (Denchak, 2017; IPCC, 2014; United Nations, 2019):

> Climate *change* is a significant variation of average weather conditions, such as, conditions becoming warmer, wetter, or drier over several decades or more. It's that longer-term trend that differentiates climate change from natural weather variability. And while "climate change" and "global warming" are often used interchangeably, global warming, the recent rise in the global average temperature near the earth's surface is just one aspect of climate change. NASA scientists have observed Earth's surface is warming, and many of the warmest years on record have happened in the past 20 years.
>
> Climate is caused when energy from the sun is reflected off the earth and back into space (mostly by clouds and ice), or when the earth's atmosphere releases energy, the planet cools. When the earth absorbs the sun's energy, or when atmospheric gases prevent heat released by the earth from radiating into space (the greenhouse effect), the planet warms. A variety of factors, both natural and human, can influence the earth's climate system.

The Anthropogenic causes of climate change is Humans, more specifically, the greenhouse gas (GHG) emissions we generate, are the leading cause of the earth's rapidly changing climate. Greenhouse gases play an important role in keeping the planet warm enough to inhabit. But the amount of these gases in our atmosphere has skyrocketed in recent decades. According to the Intergovernmental Panel on Climate Change (IPCC), concentrations of carbon dioxide, methane, and nitrous oxides "have increased to levels unprecedented in at least the last 800,000 years." Indeed, the atmosphere's share of carbon dioxide—the planet's chief climate change contributor—has risen by 40 percent since preindustrial times

Since 1970, CO_2 emissions have increased by about 90%, with emissions from fossil fuel combustion and industrial processes contributing about 78% of the total greenhouse gas emissions increase from 1970 to 2011. Agriculture, deforestation, and other land-use changes have been the second-largest contributors.

Global warming is caused by the greenhouse effect, a natural process by which the atmosphere retains some of the sun's heat, allowing the Earth to maintain the necessary conditions to host life. Without the greenhouse effect, the average temperature of the planet would be -18^{0}C.

The burning of fossil fuels like coal, oil, and gas for electricity, heat, and transportation is the primary source of human-generated emissions. A second major source is deforestation, which releases sequestered carbon into the air. It's estimated that logging, clear-cutting, fires, and other forms of forest degradation contribute up to 20 percent of global carbon emissions. Other human activities that generate air pollution include fertilizer use (a primary source of nitrous oxide emissions), livestock production (cattle, buffalo, sheep, and goats are major methane emitters), and certain industrial processes that release fluorinated gases. Activities like agriculture and road construction can change the reflectivity of the earth's surface, leading to local warming or cooling, too.

Climate change encompasses not only rising average temperatures but also extreme weather events, shifting wildlife populations and habitats, rising seas, and a range of other impacts. All of these changes are emerging as humans continue to add heat-trapping greenhouse gases to the atmosphere.

It's estimated that the earth's average temperature rose by about One (1) degree Fahrenheit during the 20th century. If that doesn't sound like much, consider this: When the last ice age ended and the northeastern United States was covered by more than 3,000 feet of ice, average temperatures were just 5 to 9 degrees cooler than they are now.

Global warming is an aspect of climate change, referring to the long-term rise of the planet's temperatures. It is caused by increased concentrations of greenhouse gases in the atmosphere, mainly from human activities such as burning fossil fuels, deforestation and farming. Global warming is an aspect of climate change, referring to the long-term rise of the planet's temperatures. It is caused by increased concentrations of greenhouse gases in the atmosphere. Human activity, such as burning fossil fuels, deforestation and farming are primary causes of concentrations of greenhouse gases

The effects of global climate change, according to the World Economic Forum's 2016 Global Risks Report, the failure to mitigate and adapt to climate change will be "the most impactful risk" facing communities worldwide in the coming decade—ahead even of weapons of mass destruction and water crises.

Blame its cascading effects: As climate change transforms global ecosystems, it affects everything from the places we live to the water we drink to the air we breathe.

Changes in climate can affect food production, water availability, wildlife and human health. Weather conditions, such as storms, can damage infrastructure like roads, rail networks and buildings. As the earth's atmosphere heats up, it collects, retains, and drops more water, changing weather patterns and making wet areas wetter and dry areas drier. Higher temperatures worsen and increase the frequency of many types of disasters, including storms, floods, heat waves, and droughts. These events can have devastating and costly consequences, jeopardizing access to clean drinking water, fueling out-of-control wildfires, damaging property, creating hazardous-material spills, polluting the air, and leading to loss of life.

Pollution is the introduction of harmful materials into the environment. These harmful materials are called pollutants. Pollutants can be natural, such as volcanic ash. They can also be created by human activity, such as trash or runoff produced by factories. Pollutants damage the quality of air, water, and land.

Air pollution and climate change are inextricably linked, with one exacerbating the other. When the earth's temperatures rise, not only does our air gets dirtier—with smog and soot levels going up—but there are also more allergenic air pollutants such as circulating mold (thanks to damp conditions from extreme weather and more floods) and pollen (due to longer, stronger pollen seasons). With anthropogenic climate change driven by human-caused emissions to the atmosphere, it stands to reason that we face compromised air quality. This affects human health, especially children. Air pollution can lead to asthma, heart and lung disease. Unchecked, according to the World Health Organization, "climate change is expected to cause approximately 250,000 additional deaths per year" between 2030 and 2050.

Since humans are the cause (stimulus) of climate change (response) because of their mindset and resulting behaviors, humans also have the capacity to change their thinking and behavior to solve the problem.

The Arctic is heating twice as fast as any other place on the planet. As its ice sheets melt into the seas, our oceans are on track to rise one to four feet higher by 2100, threatening coastal ecosystems and low-lying areas. Island nations face particular risk, as do some of the world's largest cities, including New York, Miami, Mumbai, and Sydney. Rising sea levels potentially will engulf some areas of the world, eliminating land mass.

Climate change is increasing pressure on wildlife to adapt to changing habitats—and fast. Many species are seeking out cooler climates and higher altitudes, altering seasonal behaviors, and adjusting traditional migration patterns. These shifts can fundamentally transform entire ecosystems and the intricate webs of life that depend on them. As a result, according to a 2014 IPCC climate change report, many species now face "increased extinction risk due to climate change." And one 2015 study showed that mammals, fish, birds, reptiles, and other vertebrate species are disappearing 114 times faster than they should be, a phenomenon that has been linked to climate change, pollution, and deforestation—all interconnected threats. On the flip side, milder winters and longer summers have enabled some species to thrive, including tree-killing insects that are endangering entire forests.

The earth's oceans absorb between one-quarter and one-third of our fossil fuel emissions and are now 30 percent more acidic than they were in preindustrial times. This acidification poses a serious threat to underwater life, particularly creatures with calcified shells or skeletons like oysters, clams, and coral. It can have a devastating impact on shellfisheries, as well as the fish, birds, and mammals that depend on shellfish for sustenance. Rising ocean temperatures are also altering the range and population of underwater species and contributing to coral bleaching events capable of killing entire reefs—ecosystems that support more than 25 percent of all marine life.

Trump withdraws America from the Paris Agreement

On June 1, 2017, President Donald Trump withdrew America from participation in the Paris

Agreement, delivering the following statement (White House, 2017). Trump gave the reasons for withdrawal from the Paris Agreement as Loss of American jobs and stated the agreement was at a disadvantage to America's economy, the fact remains that climate change is a clear and present threat to the world, including America, with almost certain probability to become worse if action outlined by the Paris Agreement is not taken. Nothing could be further from the truth than Trump's reasoning about the Paris Agreement. His thinking indicates his lack of ability and intellect to understand what is good for America. The loss of American jobs is because products sold in America are made outside of America. American companies moved the manufacturing outside of America and displaced workers. That is the primary cause of American jobs being loss. How many jobs will there be when people are fighting for survival?

"As President, I can put no other consideration before the well-being of American citizens. The Paris Climate Accord is simply the latest example of Washington entering into an agreement that disadvantages the United States to the exclusive benefit of other

countries, leaving American workers — who I love — and taxpayers to absorb the cost in terms of lost jobs, lower wages, shuttered factories, and vastly diminished economic production."

"Thus, as of today, the United States will cease all implementation of the non-binding Paris Accord and the draconian financial and economic burdens the agreement imposes on our country. This includes ending the implementation of the nationally determined contribution and, very importantly, the Green Climate Fund which is costing the United States a vast fortune."

"Compliance with the terms of the Paris Accord and the onerous energy restrictions it has placed on the United States could cost America as much as 2.7 million lost jobs by 2025 according to the National Economic Research Associates. This includes 440,000 fewer manufacturing jobs — not what we need — believe me, this is not what we need — including automobile jobs, and the further decimation of vital American industries on which countless communities rely. They rely for so much, and we would be giving them so little."

"The reality is that withdrawing is in America's economic interest and won't matter much to the climate. The United States, under the Trump administration, will continue to be the cleanest and most environmentally friendly country on Earth. We'll be the cleanest. We're going to have the cleanest air. We're going to have the cleanest water. We will be environmentally friendly, but we're not going to put our businesses out of work and we're not going to lose our jobs. We're going to grow; we're going to grow rapidly. "

The U.S. withdrawal from its climate change commitments in the Paris Agreement, lost time in addressing the climate change / global warming worldwide problem and threatens to bring about a period of long and dangerous warming, raising the cost and challenge to survive for future generations. Trump's withdrawal from the Paris Agreement was a setback to addressing climate change / global warming problem and was irresponsibly shortsighted, especially for someone who is the President of a country. It was irresponsible of Trump to withdraw America from the Paris agreement without an alternative plan. His decision displayed ignorance and an incompetence to lead the American people. It was plain stupid and a dangerous gamble. His domestic energy policy favors the increased production of fossil fuels. Withdrawal from the Paris Agreement only causes the world nations and our allies to have witnessed the unfit leadership and incompetence of the Trump administration. It also gives insight into the psychology of a significant faction of the American people. A significant number of Americans are gullible to believe a charlatan. "The biggest risk of all, though, is that back-sliding by the United States could quickly lock in a prolonged and dangerous warming. That would be especially likely if other nations follow Trump out of the global pact" (CUSHMAN & HIRJI, 2017; Lavelle, 2020). Trumpian influence in American politics needs to be eliminated. He needs a muffler like those placed on dogs for the survival of our society. Sometimes allowing freedom of speech is more detrimental than to take it away.

> *"This is a huge gamble," said* Nathaniel Keohane, *vice president of global climate programming at the Environmental Defense Fund. "We know the longer we wait to act, the more expensive it will be and the more drastic measures we'll have to take to avert the worst impacts of climate change. So, walking backwards on climate just when the rest of the world has committed to move forward is a huge gamble"* (CUSHMAN & HIRJI, 2017; Lavelle, 2020).

The Trump administration cut monetary contributions to the international Green Climate Fund, to assist poor countries decrease their emissions and adapt to climate changes. Trump criticized the fund as *"a redistribution of wealth that America could ill afford, calling it, a slush fund."* From the beginning, Trump had no intention of fulfilling the Paris Agreement. His focus was to dismantle practically every federal policy aimed at reducing emissions, whether under the Clean Power Plan for electric utilities or the many other initiatives of the Obama administration. One has to wonder whether Trump's withdrawal from the Paris Agreement was reflective of his ignorance and unfit ability to lead the American people or whether is was a prejudicial jealousy toward previous President Barack Obama.

Withdrawal by the U.S. could slow the process of switching to cleaner energy, especially for the smaller and poorer countries. The point of diminishing returns is only a few years away, according to scientist. Republican politicians have a counterproductive view to that of scientific opinions. In a worst-case scenario, the U.S. abandonment of its commitment, could warm the earth an additional .03 degrees Celsius by the end of the century. Already U. S. withdrawal has offset planned emissions reduction of 28% by 2020 from 2005 levels (CUSHMAN & HIRJI, 2017; Lavelle, 2020). The U.S. is one of the biggest greenhouse polluters of all and it is irresponsible to leave the rest of the world to compensate for America's irresponsible neglect.

China, the world's top emitter, and India, the third-highest emitter of climate pollution, are on track to meet their climate pledges a few years ahead of schedule, the politics of America are narrow minded and alienating from the rest of the world. European and Chinese diplomats and UN leaders have spoken forcefully in favor of preserving the Paris Agreement and vowed to stay the course, no matter what Trump does. They regard the Paris pact as a "historic achievement" and "irreversible" (CUSHMAN & HIRJI, 2017; Lavelle, 2020).

> *"The nations that remain in the Paris Agreement will be the nations that reap the benefits in jobs and industries created. I believe*

*the United States should be at the front of the pack. But even in the absence of American leadership," he said, "I'm confident that our states, cities, and businesses will step up and do even more to lead the way" (*CUSHMAN *&* HIRJI, *2017;* Lavelle, *2020).*

Trump's racism, jealousy and envy of President Barack Obama is hypothesized as a reason Trump withdrew the United States from the Paris Agreement. Whether the fact that President Obama being a Black man instigated Trump's alleged racism, is raised as another issue. The fact that Donald Trump would allow personal issues to enter his decision-making for the American people raises a serious issue regarding his qualification as a leader. The Paris Agreement and the Iran Nuclear Agreement were landmark international achievements of the Obama administration. Along with the Affordable Health Care Act. President Obama's leadership delivering America out of a potential financial depression displayed good leadership and a patriotic concern for America. These are milestone achievements of the Obama administration. Allegedly, Trump's racism and jealous envy motivates him to undo President Obama's achievements, at the expense of the people of the United States of America and safety of the world.

The perfect supporting cast member of Donald Trump's irresponsible anti-climate warming agenda is Mike Pence. Vice President Pence's record on climate is built on support for oil, gas, and coal, his dedication to the industry and strong opposition to the Paris climate agreement compliments Trump's political agenda. There is more background to Pence being selected as Trump's Vice President. In addition to Pence being a "yes" man for Trump. Vice President Mike Pence's intentions to thwart climate action began long before he became President Donald Trump's Vice President. The former Indiana governor came into Trump's orbit with a reputation as a culture warrior who sought to restrict gay marriage and reproductive rights, and he has described his defense of the fossil-fueled economy as part of the same ideological battle. The flaw in Pence's viewpoint is that he expresses his views in terms of politics instead of showing concern for the best interest of people. Pence is a politician bought and owned by the fossil fuel industry, a worthless politician when it comes to serving the interest of the American people.

> *"It has long been a goal of the liberal left in this country to advance a climate change agenda,"* Pence said in a June 2017 interview on Fox & Friends in support of Trump's decision to exit the Paris climate accord.
>
> *"For some reason or another, this issue of climate change has emerged as a paramount issue for the left in this country and around the world."* (Lavelle, 2020).

The Koch brothers, owners of Koch Industries, is one of the major polluters in America (Greenpeace, 2022). The Koch brothers gave $300,000 toward Pence's run for Indiana governor in 2012. As governor of Indiana, a state second to Texas in coal consumption, Pence could influence national climate policy. The previous Republican governor, Mitch Daniels, had established a renewable energy program and started an energy efficiency program for Indiana.

> *"Another key asset Pence brought to the Trump administration was the close bond he had forged with wealthy corporate donors who shared his determination to prevent U.S. action on climate, especially the petrochemical, libertarian, billionaire Koch brothers"* (Lavelle, 2020).

But with Pence as governor, the Republican legislature dismantled those programs (Lavelle, 2020). In 2014 Indiana lawmakers passed a repeal of the energy efficiency initiative. Pence said he had mixed feelings, but killed the program decisively when he neither signed nor vetoed the legislation, allowing the repeal to become law in 2015 without his signature. In Pence's 2015 State of Indiana address, he declared an all-out war against the Obama administration's Clean Power Plan, a climate change initiative aimed at reducing carbon emissions from coal plants. "Indiana is a pro-coal state," Pence said. "We must continue to oppose the overreaching schemes of the EPA until we bring their war on coal to an end" (Lavelle, 2020). Pence became an asset for Koch Industries, and likewise antagonist to Clean Air Programs. The Trump administration eventually repealed the Clean Power Programs.

The Impact of Climate Change on the Environment

In 2018, The Fourth National Climate Assessment concluded that the impacts of global climate change have affected the US. Clear empirical evidence supports the conclusions based on the growing number of intense tornadoes, hurricanes, earthquakes, wildfires, droughts, heatwaves, and floods that the country has been experiencing. These impacts will only get worse in the future unless urgent action is taken to curb greenhouse gas emissions. It notes that annual average temperatures across the US have risen by 1.8°F since the beginning of the twentieth century. The Fourth National Climate Assessment (NCA4), completed in November 2018, is a comprehensive and authoritative report on climate change and its impacts on the United States (MITRA, 2018; Globalchange.gov. 2018). The 1,600-page report directly connects climate change to natural disasters impacting human lives and resources in the US. Wildfires in California, declining water levels in the Colorado River Basin, and the spread Lyme disease and other vector-borne diseases like West Nile and Zika. Every area of the country is threatened by the impacts of a global warming world. These impacts are felt more intensely by the most vulnerable, children, the elderly, the poor, and communities of color.

The report notes that climate-related disasters have already cost the US billions of dollars every year in economic losses. For example, rebuilding Puerto Rico's power grid, which was destroyed by hurricanes Irma and Maria, is estimated to cost $17 billion. Flooding in the Mississippi and Missouri river basins in 2011, triggered by exceptionally heavy rainfall, caused an estimated $5.7 billion in damages. The federal cost of fighting fires across the US ranged from $809 million to $2.1 billion per year between 2000 and 2016. By 2100, these costs could rise to hundreds of billions of dollars a year and could affect the US economy worse than the Great Recession did, according to the report (MITRA, 2018; Globalchange.gov. 2018).

"The assessment, which is required by law and is released in installments over four years, was compiled by the US Global Change Research Program, a consortium of 13 federal agencies including the Department of Defense (DOD), the Environmental Protection Agency (EPA), the National Aeronautics and Space Administration (NASA), as well as a group of independent scientists from across the country. It's the second of two volumes. The first, released in November 2017, concluded that human activity, "*especially emissions of greenhouse gases,*" was the key driver of climate change" (MITRA, 2018; Globalchange.gov. 2018). In the face of this report and data compiled by U. S. government agencies, the Trump administration, Donald Trump and many Republicans take the position they do not "believe" the findings. Their disbelief is based on nothing they have done to verify their position. This is an example of incompetence, denial based on ignorance and a disgrace to human intelligence. The report further states: *"Americans increasingly recognize the risks climate change poses to their everyday lives and livelihoods and are beginning to respond."* What Trump and his cohorts fail to acknowledge is that silly tweets and press statements downplaying the real impact and causes of climate change to mislead the public, only magnify their ignorance. Ultimately, the truth will come out, by the lived experiences of constant disasters and devastation people will encounter. If we don't wake up to the urgency of this crisis, disaster is on the horizon.

If the global temperature rises by 1.5°C, humans will face unprecedented climate-related risks and weather events. The image to the right indicates the environment response to a temperature rise of 1.5 degrees Celsius and 2 degrees Celsius. We are currently on track for a 3-4°C temperature rise. One half a degree temperature increase may be the difference between a world with coral reefs and Arctic summer sea ice, and a world without them (Cool earth, 2018).

The important thing to take away from this section of the book are:

1. The earth is in is experiencing climate temperature increase
2. Humans are in grave danger of having their way of life changed drastically if climate change is not controlled.
3. Climate management requires immediate international action by all countries to be controlled.
4. If the earth's temperature rises by 1.5 degrees or 2 degrees Celsius, wildfires, earthquakes, floods, hurricanes, and tornadoes will increase. Food supplies will be threatened and decrease. Fresh water supplies will be threatened.

5. We are currently on track for a 3-4°C temperature rise within the next 10 years. If that happens the climate conditions on earth will be life threatening.

6. Addressing climate change is critical. Immediate action is required. Once the environmental conditions become life threatening, human intervention to reverse the situation will be a huge problem, if not impossible.

7. Human actions are the primary cause of climate change and global warming.

Global Warmers / Fossil Fuel Emitters

Research finds that 100 active fossil fuel producers; including ExxonMobil, Shell, BHP Billiton, Koch Industries and Gazprom are responsible for over 70% of industrial greenhouse gas emissions since 1988, the same year the Intergovernmental Panel on Climate Change (IPCC) was established (CDP, 2017). The following facts are also worth knowing (CDP, 2017):

Almost a third (32%) of historic emissions come from publicly listed investor-owned companies, 59% from state-owned companies, and 9% from private investment

Over half of global industrial emissions since 1988 can be traced to just 25 corporate and state producers

Fossil fuel companies and their products have released more emissions in the last 28 years than in the 237 years prior to 1988

Over half (52%) of all global industrial GHGs emitted since the start of the industrial revolution in 1751, have been traced to these 100 fossil fuel producers

Climate change can be mitigated if urgent action is taken

The 100 fossil fuel producers are the source of 635 billion tons of GHGs emitted since 1988, the year human-induced climate change was officially recognized. The data also shows that 32% of these emissions come from companies that are public investor-owned, highlighting the power of investors in the transition to a sustainable economy. The dilemma for investors is how to save the planet and still gain a return on their investment. Global fossil fuel emissions are concentrated over a small number of producers. From 1988 to 2015, just 25 fossil fuel producers are linked to 51% of global industrial GHG emissions. The highest emitting companies over the period since 1988 include: public investor-owned companies such as ExxonMobil, Shell, BP, Chevron, Peabody, Total, and BHP Billiton. State-owned entities such as Saudi Aramco, Gazprom, National Iranian Oil, Coal India, Pemex, CNPC and Chinese coal, of which Shenhua Group & China National Coal Group are likewise major fossil fuel contributors (CDP, 2017).

Biden Reinstates America into the Paris Agreement

The Presidential election of 2020 resulted in a defeat for Donald Trump (R) and a victory for Joe Biden (D). One of the first acts of Joe Biden's Presidency was to reinstate America back into the Paris Accord Agreement. On January 20, on his first day in office, President Biden signed the instrument to bring the United States back into the Paris Agreement. Per the terms of the Agreement, the United States officially becomes a part of the agreement again. One of the disappointments in Donald Trump's withdrawal from the Paris Agreement is that when the Paris Agreement was installed in 2016, under the Obama administration, America was one of the key sponsors and leading entities in constructing the Agreement between nations. Withdrawal by America was comparable to reneging on a promise (Blinken, 2021), or comparable to America withdrawal from World War II, or comparable to George Washington quitting the Revolutionary War. Almost 200 nations signed on to the Paris Agreement and committed to limit their greenhouse gas emissions to keep global warming below 2 degrees Celsius - preferably below 1.5 degrees Celsius - compared to pre-industrial temperatures.

The dynamics of America's inclusion, exclusion and inclusion in the Paris Agreement was a clear signal that Republicans and Democrats have opposing agendas. Much time and momentum was lost in the construction of a firm and continuous American energy policy during the Trump years in office. The significance of America's participation in the Paris Agreement and climate control policies is evident in the 2020 climate activity. There were 16 climate-driven disasters that cost at least $1 billion each, according to the National Oceanic and Atmospheric Administration. No other country has emitted more cumulative carbon dioxide into the atmosphere since the industrial era began in the mid-1800s than the United States. And even though U.S. emissions are falling, the rate of the reduction is far too slow to avoid catastrophic warming, according to climate scientists (Mai, 2021).

One weakness in the Climate policy of the United States is that as politics and leadership in America changes and bounces from party to party, Democrat to Republican, Republican to Democrat. The climate policy needs continuous and stable bi-partisan support. Without bi-partisan continuous support, America will appear to be a "fickle" and unstable country, resulting in a diminished world-wide respect.

Chapter III

The Role of Fossil Fuels in Global Warming/Climate Change

Over the years mining, drilling and the burning of, what is referred to as dirty energy (fossil fuels), causes harm to our environment and health. The burning of fossil fuel's has been the primary source of energy in the World for over a century. Fossil fuels power our cars, businesses, and supplies energy to our homes. In our contemporary society, fossil fuels supply approximately 80% of our energy requirements. However, the comfort provided by the burning of fossil fuels does not come without a price. We are not only talking about the dollar cost of using fossil fuel energy but the cost to humanity and cost at the expense of the environment. There is growing concern that the price for comfort is too high a price to pay when our survival as a human species is at stake.

What are fossil fuels? Fossil fuels are coal, crude oil (petroleum) and natural gas. They are called fossil fuels because they were formed from fossilized, buried remains of plants and animals that existed millions of years ago. It is the high carbon content of fossil fuels that cause environmental pollution (Denechak, 2018).

The image to the left shows land degradation that is caused by excavating for fossil fuel resources. The installation of wells, pipelines, access roads, storage facilities for raw materials, waste storage facilities are examples of how the recovery of fossil fuels from the

earth has a stripping effect. It is called "strip mining" because the land is stripped of its natural habitat causing the natural texture of the land to be destroyed and the natural vegetation, animal, and wildlife habitation to be fragmented and destroyed. The mining of fossil fuels is a cause of

air pollution and causes pollution to clean air and water (Denechak, 2018.)

The burning of fossil fuels causes emissions into the environment and air pollution that can cause breathing problems for humans. A major cause of air pollution are factories in the production of goods, consumer, and industrial products.

Air pollution causes problems for humans, especially those with breathing and respiratory conditions. A major cause of air pollution are

factories in the production of goods, consumer, and industrial products.

Transportation is a major source of air pollution. Cars trucks and buses that are powered by fossil fuels. (gasoline and diesel fuel). These gas and diesel fueled vehicles contribute more than one half of the nitrogen oxides into the environment. Automotive vehicles are a primary source of global warming (Union of Concerned Scientist, 2018).

Fetuses, newborn children, and people with chronic illnesses are especially susceptible to the effects of air pollutants. Pollution from motor vehicles is identified in the following categories (Union of Concerned Scientist, 2018):

- Particulate Matter (PM) is the soot from the vehicle exhaust.
- Volatile Organic Compounds (VOC's) react with nitrogen oxides when exposed to sunlight and forms ground level ozone. A primary cause of smog and lung disease.
- Nitrogen oxides (Nox) form a ground level ozone that can cause lung irritation and weaken the body's immune system.

- Carbon Monoxide (CO2) is an odorless, colorless, and poisonous gas emitted from cars and trucks, which if inhaled can be deadly.
- Sulfur dioxide (SO2) emitted mainly by power plants from burning sulfur dioxide. A significant risk to children and asthmatics.
- Greenhouse gases are carbon dioxide pollutants that contribute to global climate change. Emitted from (airplanes, trains, ships, cars and trucks.

The silver lining in the case of vehicle emissions is that automobile manufacturers are giving significant effort into making energy efficient electric vehicles that will operate on clean electric energy.

The following indicates the sources of fossil fuel emissions (CDP, 2017):

> One hundred fossil fuel producers including ExxonMobil, Shell, BHP Billiton and Gazprom are responsible for 71% of the industrial greenhouse gas (GHG) emissions since 1988, the year in which human-induced climate change was officially recognized through the establishment of the Intergovernmental Panel on Climate Change (IPCC).

Almost a third (32%) of historic emissions come from publicly listed investor-owned companies, 59% from state-owned companies, and 9% from private investment.

> Over half of global industrial emissions since 1988 can be traced to just 25 corporate and state producers.
>
> Fossil fuel companies and their products have released more emissions in the last 28 years than in the 237 years prior to 1988.

Over half (52%) of all global industrial GHGs emitted since the start of the industrial revolution in 1751, have been traced to these 100 fossil fuel producers.

A Low carbon tipping point is possible to achieve if investors and carbon majors take urgent climate action.

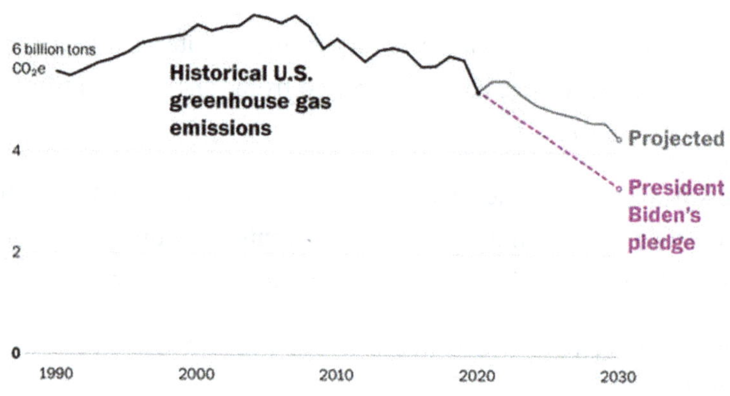

Note: Chart shows the center of a range of projected emissions under current U.S. policies

Source: U.N. Framework Convention on Climate Change, Rhodium Group

As of the 2021, the US goal for reducing the GHG effect is seriously falling short, and for each month that passes. It becomes harder to succeed, until at some point soon, it will become virtually impossible. That's true for the United States, and also true for the planet, as nearly 200 nations strive to tackle climate change with a fast-dwindling timeline for doing so.

Joe Manchin (D. W-VA) is an obstacle to the environmental crisis. According to him it is because addressing the climate problem will cause inflation (Mooney & Stevens, 2022). Now how stupid is that response from Manchin. Inflation compared to the end of human existence. His logic is illogical. However we all know the real reason Manchin is an obstacle to the climate change / global warming agenda (Brady & Maxine, 2022). In order to find out why Manchin is opposing his own party regarding climate change / global warming policies, we need to look into his background, being from West Virginia.

Manchin founded the coal brokerage Enersystems in 1988, and helped run it until he became a full-time politician. When he was elected West Virginia secretary of state in 2000, he gave control of Enersystems to his son Joseph. Manchin is financially connected to the fossil fuel industry. His argument against the agenda for global warming / climate change is not mediate inflation for the American people. His argument is personal based on financial gain, for himself and his family (ENERSYSTEMS, INC., 2023; Lewis, 2022; Schouten, 2022).

Chapter IV

The Relationship between Fossil Fuel Emissions, Industrialization, and the Environment

The Stimulus

There is no doubt that the Industrial Revolution that transitioned the manufacturing processes in Europe and the United States had a serious impact that laid the foundation for the development of modern business. The Industrial revolution took form and created a stabilizing impact from approximately 1760 to 1840. This development in European and American history produced several benefits for consumers, our American and World Society. The industrial revolution can be summarized as the period where the manufacturing of goods moved from homes and small shops into large factories, a shift that resulted in cultural transformation and migration of people from rural areas to cities. Over time factories developed machinery that made the manufacturing of goods more efficient. However, over time, this development has resulted in a counterproductive result upon the environment. The increase of industrialization over the years has produced more carbon into the environment, releasing carbon dioxide into the atmosphere by the burning of fossil fuels (oils, gas, coal). The fossil fuels released into the environment has change the world's climate (Exploratorium, 2020).

Carbon dioxide released into the Environment over the last 380 years has increased significantly (Catalina, 2020). This chart shows that the use of fossil fuels increased from 270 units of measure in 1750 to almost 350 units of measure in 2000.

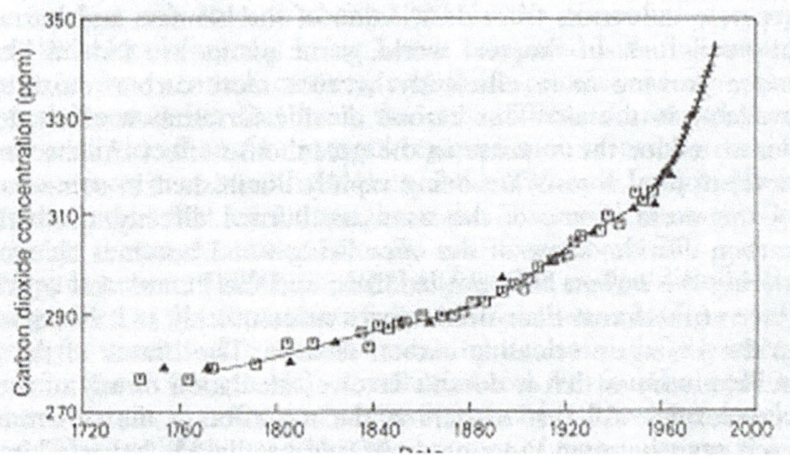

Human population increased over the last 300 years and fossil fuel usage increased over the past 150 years, significantly (Catalina, 2020). The chart below shows that the World human population increased from 500 million in 1700 to over 3 billion in 2000. In the year 2019, the world global population is estimated to be 7,577,130,400 (7.7 billion) people

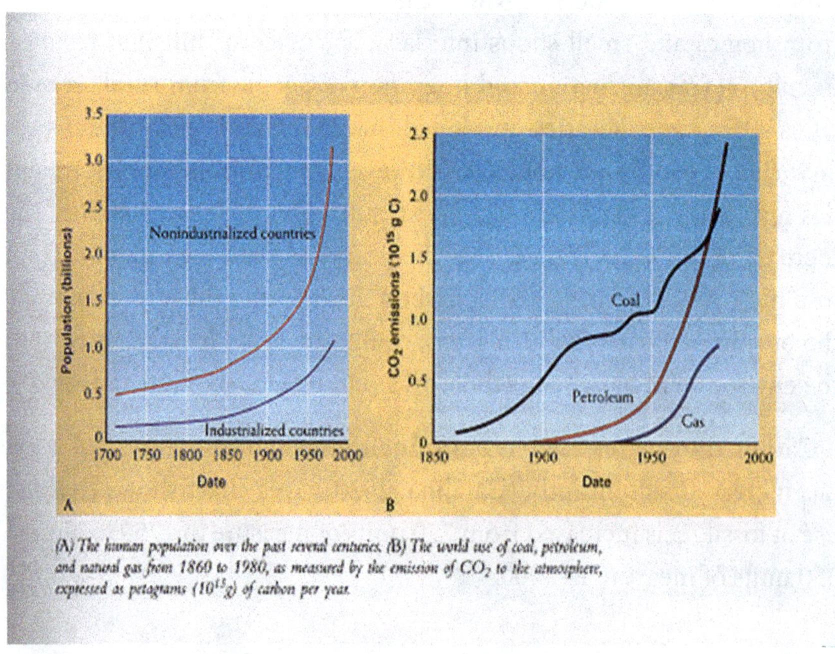

(A) The human population over the past several centuries. (B) The world use of coal, petroleum, and natural gas from 1860 to 1980, as measured by the emission of CO_2 to the atmosphere, expressed as petagrams (10^{15}g) of carbon per year.

(World Population Prospects. 2019 Revision). Likewise, there was a rise in fossil fuel emissions because of an increase in the demand for goods.

With the growth and development of industrialism and machinery, it's become evident that humans can shape and change the environment. The ice age that ended 10 thousand years ago should make humans aware that the earth is not invulnerable to drastic change and the destruction of organic life (Catalina, 2020). Naturally, the use of fossil fuels increases with the increase in population. The demand for goods increases with population growth.

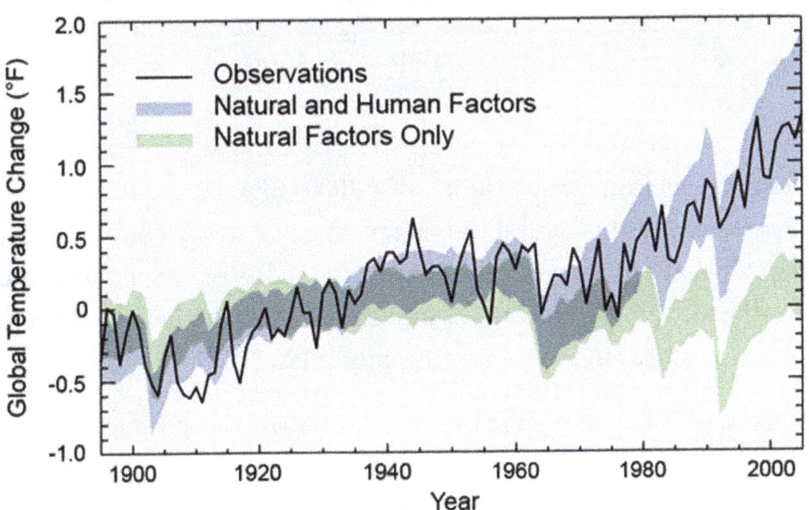

The chart below shows that over the years from 1900 to 2000 the human factors influencing the environmental have increased while the natural environmental influencing the environmental change have remained relatively stable. The natural factors seem to have a synchronization with the environment that is consistent, while he human factors that influence the environment is significantly greater (EPA, 2017).

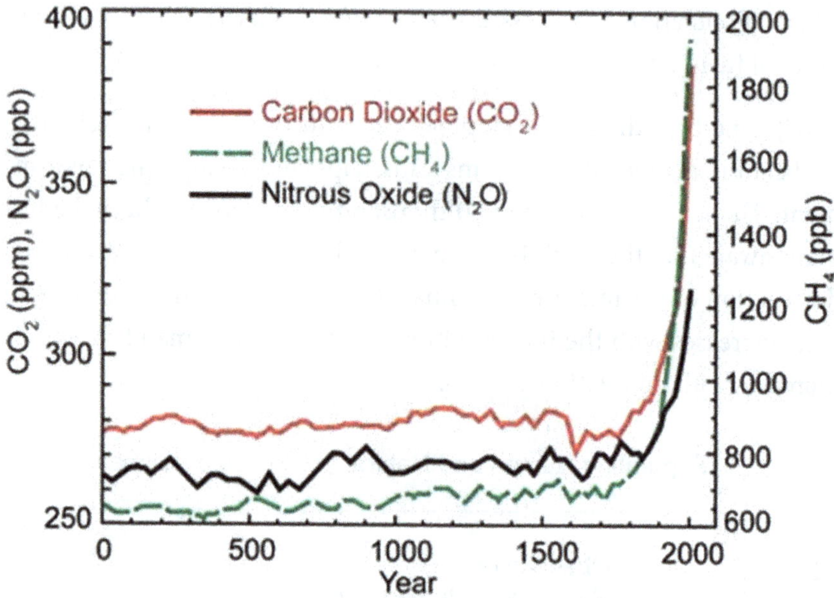

The chart below shows how the increase of fossils fuels has grown over the years from years, 0 to the year 2000 (EPA, 2017). The fossil fuel emission caused by natural causes remained relatively consistent until approximately 1700, about the time of the industrial revolution. Then coordinate with the increase of industrialization the increase in fossil fuel emission has increased significantly (EPA, 2017).

The chart below shows the Industrial countries that produce the most fossil fuel emission and the counties that emit the most fossil fuels (Mgbemene, Nnaji, & Nwozor, 2016). The leader of fossil fuel emission is America.

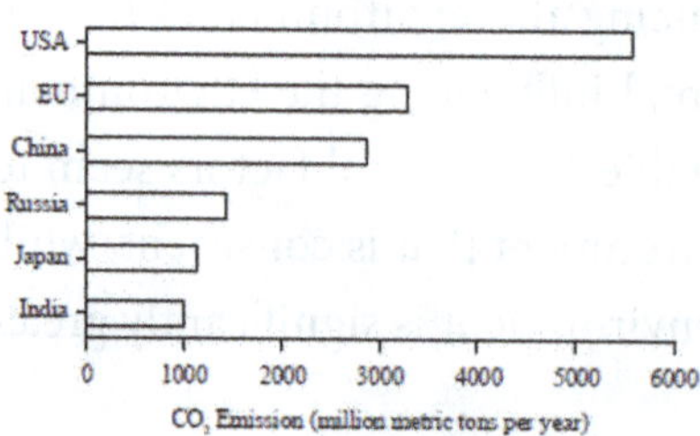

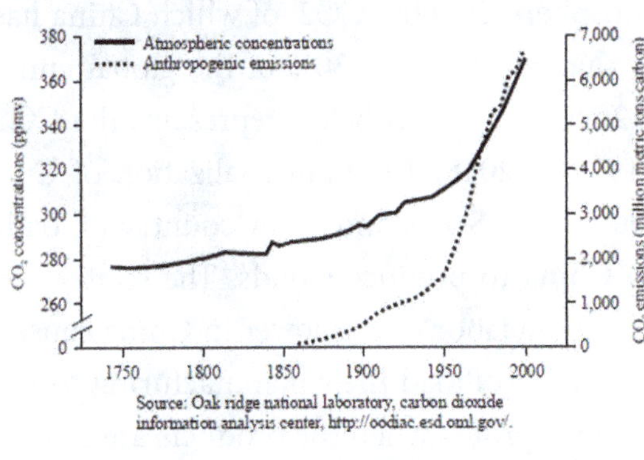

The chart on the left indicates the concentration of fossil fuel emissions in the atmosphere and the anthropogenic emissions refers to the amount of the fossil fuel emission from human activity (mostly from environmental pollution and pollutants) (Mgbemene, Nnaji, & Nwozor, 2016). It should be noted that the amounts of emissions increased significantly during the beginning and ongoing era of the industrial revolution and industrialization growth.

Geophysics is a subject of natural science concerned with the physical processes and physical properties of the Earth and its surrounding space environment, and the use of quantitative methods for their analysis (BGA, n.d). The relationship between the earth's land mass and the earth's atmosphere has been so imbalanced by human activity that climate related natural disasters have increased over the years, along with the cost of disaster damage.

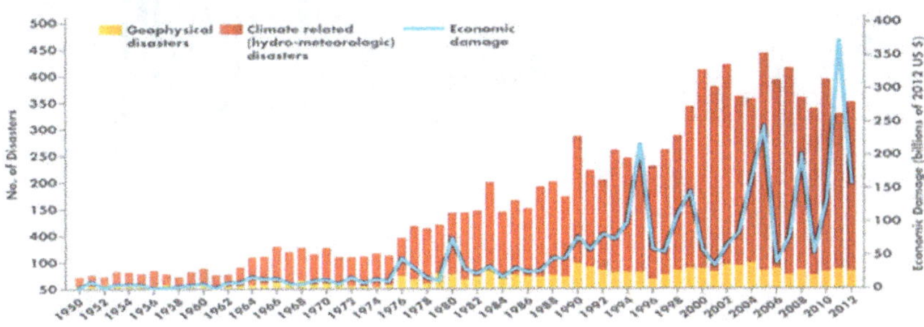

The main countries that pollute the environment are the United States, Europe and China. Cars and manufacturing account for most of the pollutants in the atmosphere. It is the CO2, of which China has become the largest emitter that accounts for 30% of the global emissions (Xiaoqi Zheng et al., 2020). The chart below represents the CO2 emissions in China from 1978 to 2018. The industrialization of China has congruence with the United States and other country offloading their manufacturing to China to produce Goods. The strategy to market cheaper goods because the labor cost is lower in China caused many manufacturers worldwide to offload their manufacturing to the Republic of China to gain higher profits when the products are sold in their home country. Products made outside of American and sold in America for higher profits, to displaced workers who are the victims of offshore manufacturing, motivate air pollution activity.

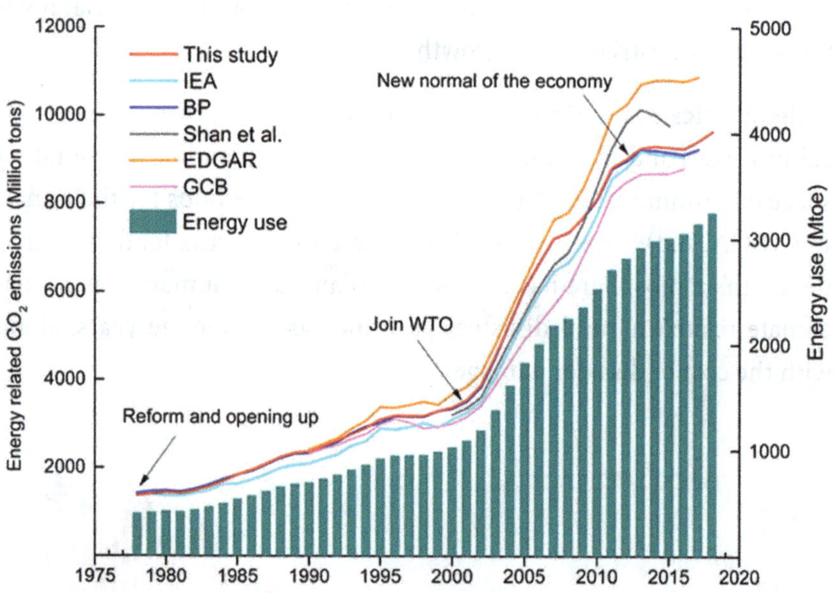

Chapter V

The Environmental Response to Climate Change caused by fossil fuel emissions:

The Response to Stimulus

Farmland flooding reduces agricultural production

(Mgbemene, Nnaji, & Nwozor, 2016, p. 26).

Air pollution is considered the stimulus that causes the environment to respond with climate change and an increasing onslaught of natural disasters. The chart above indicates an increase in natural disasters from 1952 to 2009. The incidence of natural disaster has risen dramatically over the last 60 years. To close observers of current events in relation

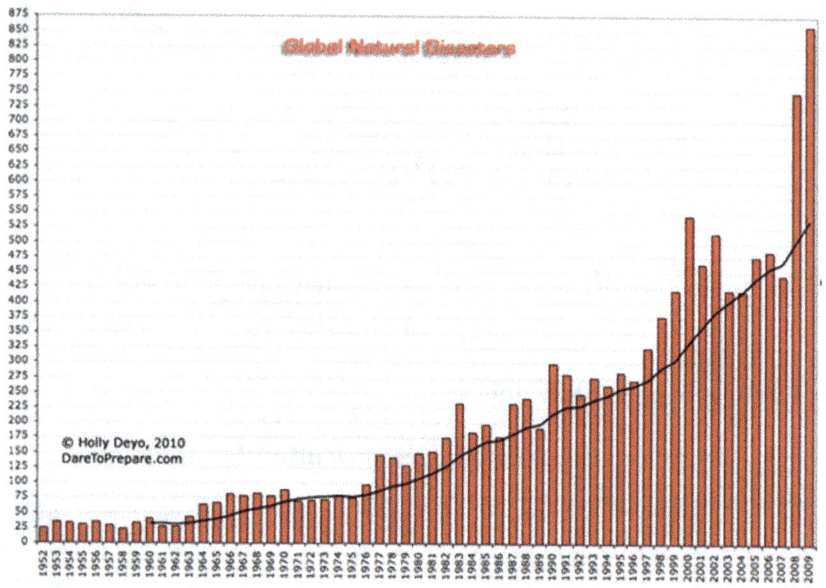

to both history and Bible prophecy, this is no mere coincidence. What muddies the water as soon as Bible prophecy is mentioned in relation to natural disaster, is the fact that there is a literal abundance of kooks, screwballs, and fanatics out there who instantly seize on the latest catastrophe to declare "*the end is nigh.*" Doomsayers have been doing this for years. Far too many get involved in such activity for personal profit. They peddle their scaremongering by taking advantage of vulnerable people and, for a price, promise to show a way that the gullible can escape the wrath of their god, the next disaster, creatures from outer space, the end of the world, the implosion of the universe or some other such wacky idea. They make it doubly hard for the genuine seeker after truth (Fraser, 2010).

There has been a significant increase in natural disasters over the past 15 years. The Center for Research on the Epidemiology of Disasters (CRED) has collected data showing that, from the period of 2000 to 2009 there were 385 disasters, which shows an increase of 233 percent over the period from 1980 to 1989 and a 67 percent increase from 1900 to 1999. Earthquakes are responsible for 60 percent of the natural disasters from 2000 to 2009. Overall, disasters related to climate, droughts, storms, floods have increased tenfold since researchers began collecting data in 1950 (Sina, 2010; CRED, 2018). While critics have responded that CRED did not present clear evidence that the disasters were a result of climate change, correlating the related factors and phenomena of industrial growth, emission of fossil fuels into the environment and natural disasters increased in relationship to the years these factors developed, there is a close alignment. It does not take a genius to see the correlation between these factors.

Increase in Natural Disasters

Incidents of natural disasters have risen considerable over the past 50 years as indicated in the chart on Global Natural Disasters. Naively, some observers claim the natural disasters are Bible prophesy. However, it does not take much intelligence to predict disaster if one observes the actions of people responsible for polluting the environment by dumping toxic waste into the air, water streams, lakes, rivers, and ocean. According to CRED statistics 3,852 natural disasters have killed more than 780,000 people over the years from 2000 to 2009. Over 10 billion people were disastrously affected during this period, at a cost of over $960 billion. Earthquakes took the highest toll (60%), disastrous storms were next (22%), and then extreme temperatures (11%) (Fraser, 2010; CRED, 2018). While some people want to give reason for the increasing world disasters to biblical prophecy about the world reaching a point in time when catastrophic events will occur because the creator will come to intervene in human affairs. He will no longer tolerate human immorality (Fraser, 2010). This viewpoint is only another way to describe the universe reaction to S-R) theory. As humans continue to act in ways that are detrimental to our earth, the earth naturally responds to human disregard for the environment. Framing the discussion in terms of biblical prophecy gives the impression of fanaticism and irrational thinking. However, to scientifically and logically present the results observed and with evidence presented in a logical trustworthy manner, makes more sense and gives believability to the situation. Only an idiot can disregard logical data supported by evidence and correlated results.

Tropical Cyclone Climatology (Schreck III, Knapp and Kossin, 2014)

- **WMO** – Climatology based on reports from the World Meteorological Organization (WMO) official forecast centers (referred to as RSMCs and TCWCs).
- **NHC+JTWC** – Climatology based on the combination of the National Hurricane Center (NHC) best track data for the North Atlantic and Eastern Pacific the Joint Typhoon Warning Center (JTWC) for other basins.

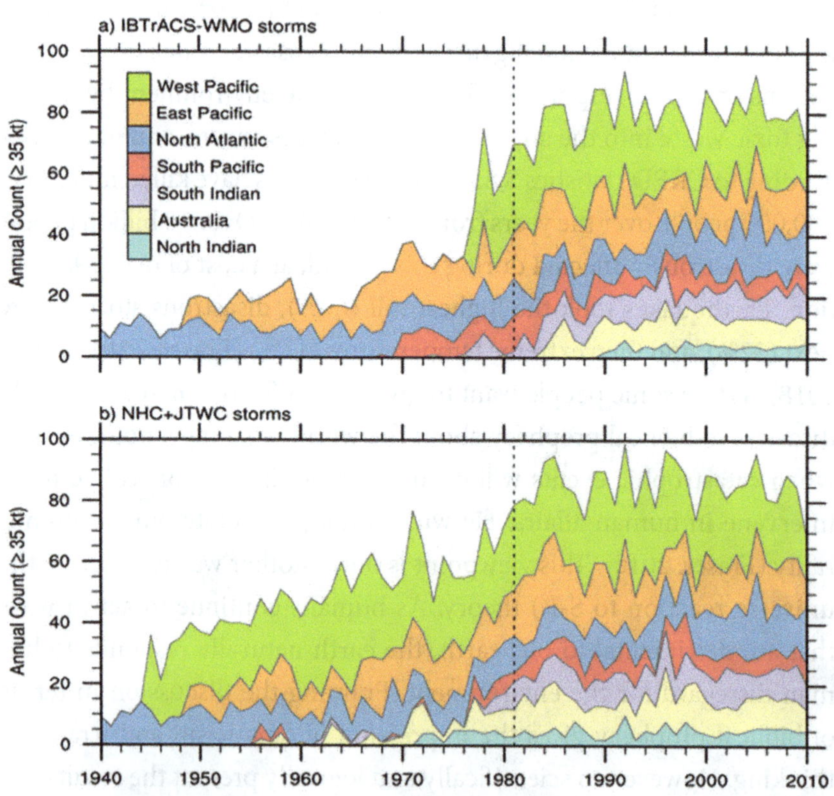

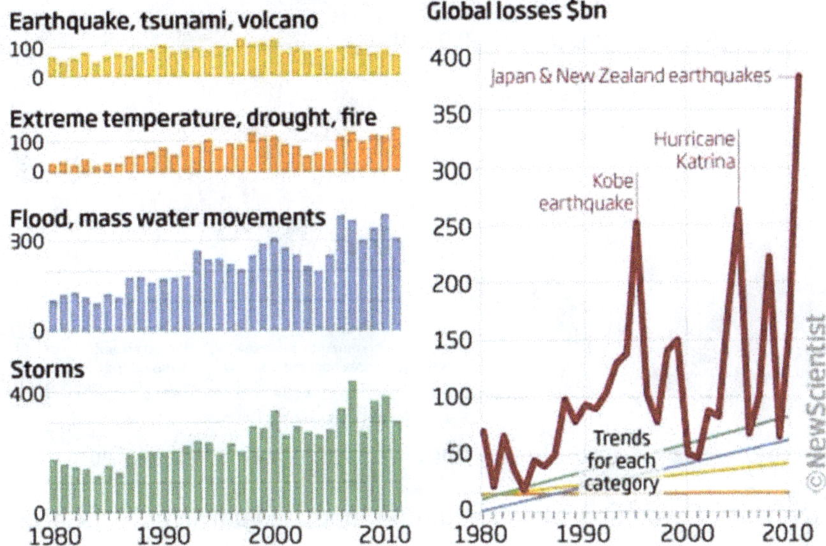

For the most part, increases in global storms has increased since 1940 in most of the regions of the world, West Pacific, East Pacific, North Pacific, moderately in the South Pacific, and the South Indian Ocean region. Only the Australian and North Indian regions have maintained a lower-level incident of disaster storm incidents.

> *"2018 was the fourth warmest year on Earth since 1850. Global mean temperature in 2018 was colder than 2015, 2016, and 2017, but warmer than every previously observed year prior to 2015. Consequently, 2016 remains the warmest year in the period of historical observations. The slight decline in 2018 is likely to reflect short-term natural variability, but the overall pattern remains consistent with a long-term trend towards global warming" (Berkeley Earth, 2020).*

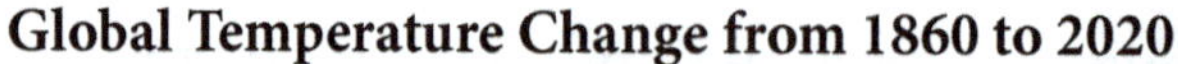

Global Temperature Change from 1860 to 2020

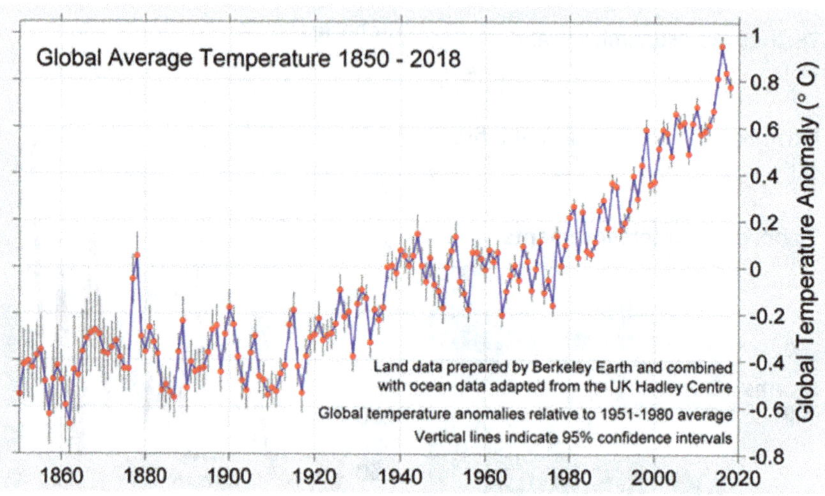

Not only have the earth temperatures increased but so have the ocean temperatures (represented in blue) (Berkeley Earth, 2020).

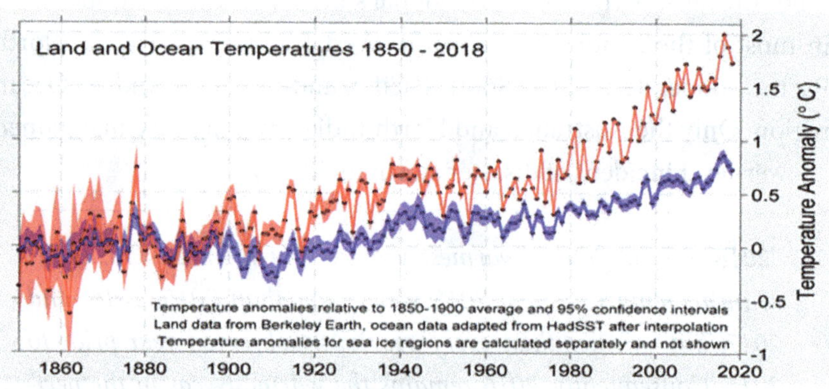

The land on average warms more than the ocean. In 2018, there was considerable warming over the Artic region. Scientist found the Artic warming exceeded the "mean ratio" warming or the earth. Warming of the Artic region and the melting of sea ice caused more sunlight to be absorbed by the ocean, which allowed for yet more warming. 2018 was the sixth warmest year in the Arctic. Arctic and earth warming was expected by scientist based on the increases in the greenhouse gas concentrations and amount of Carbon dioxide released in the environment. 2018 saw new records for both the level of carbon dioxide in the atmosphere and the annual amount of carbon dioxide emitted by human activities (Berkeley Earth, 2020). Consistent with the research of Berkeley Earth (2020) is the Climate Change map below produced by NASA (NASA, 2020), which shows significant increase in global temperatures over the last 40 years from 1980 to 2020

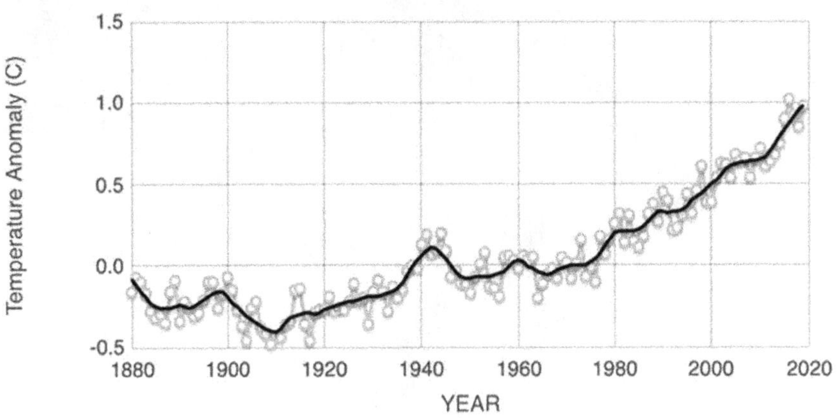

Source: climate.nasa.gov

Chapter VI

Impacts of Climate Change on humans: The Human Effect

Man is the ultimate recipient of the adverse effects of **climate change**. The world development report from the World Bank 2010 summarizes the most important consequences of climate change above 2°C as follows (Mgbemene, Nnaji, & Nwozor, 2016):

Significant loss from the Greenland and West Antarctic ice sheets and subsequent sea-level rise

Increase of floods, droughts, and forest fires in many regions

Increase of death and illness from the spread of infectious and diarrheal diseases and from extreme heat

Extinction of more than a quarter of all known species and

Significant declines in global food production

Natural disasters kill approximately 60,000 people a year and are responsible for 0.1% of global deaths. Based on a world population of 7 billion people, natural disasters have killed 700,000,000 (700 million) people (Ritchie & Roser, 2014). Previously the most fatal disasters to people were droughts and floods, today deaths from these disasters are low. The deadliest events today are earthquakes.

Some areas of the world have experienced more devastation from disasters than others.

Those at low income and impoverished areas are often the most vulnerable to disaster events: improving living standards, infrastructure and response systems in these regions will be key to preventing deaths from natural disasters in the coming decades (Ritchie & Roser, 2014).

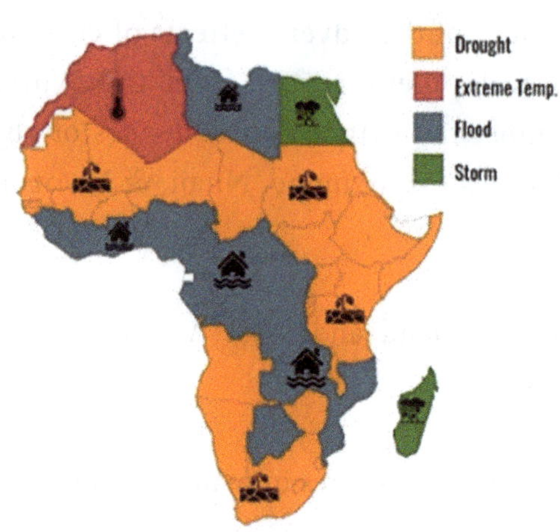

For example, the entire continent of Africa has been affected by several different disasters depending on the geographic location of the African country (see chart to the left). Poorer regions of the world and low incomes populations are usually the most vulnerable to disasters because they lack the proper living standards to protect them. The infrastructure and warning response systems in these poorer areas are not sufficient to give notice and forewarning of disaster. Upgrading the infrastructures and warning prevention systems in these areas will be key to preventing deaths from natural disasters in the coming decades. That is unless authorities continue to minimize the impact of global warming and deny the crisis we potentially face as a world. In 2010, 70% of the deaths in Haiti were due to the Port-au-Prince earthquake. The country's infra-structure was not strong enough to withstand the devastating impact of the earthquake.

Over the years, global deaths from disasters have declined, however the natural disasters have increased (Sina, 2010; CRED, 2018). Disasters include earthquakes, volcanoes, landslides, famines and drought, hurricanes, tornadoes, cyclones, Tsunamis, flooding, extreme heat (hot and cold), storms, flooding, wildfires, Because of better identification and reporting of disasters, giving people early warning they can take precautions prior to the disaster striking their environment. This is a good improvement over the years; however, the real danger is that the disasters, themselves, have become more frequent (see chart below).

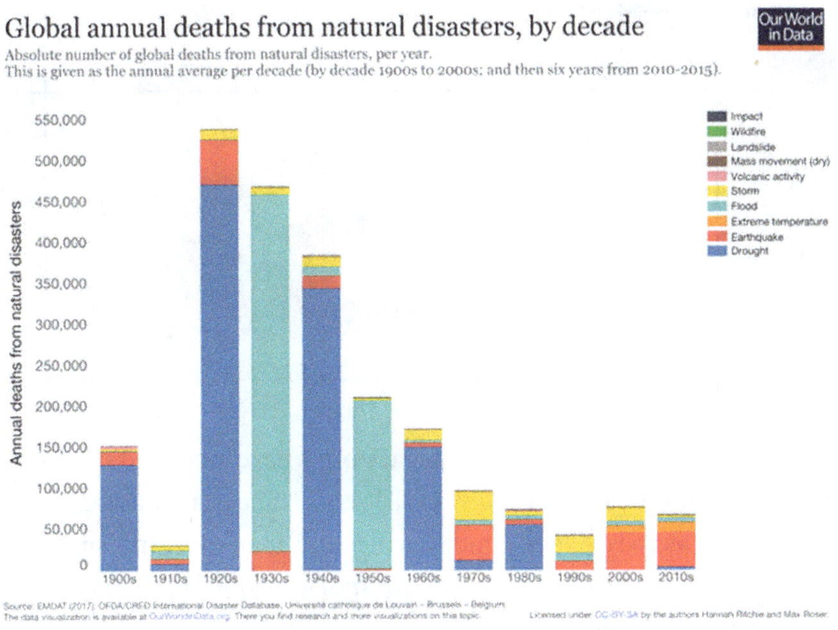

Tropical Cyclone Climatology (Schreck III, Knapp and Kossin, 2014)

a) **WMO** – Climatology based on reports from the World Meteorological Organization (WMO) official forecast centers (referred to as RSMCs and TCWCs).

b) **NHC+JTWC** – Climatology based on the combination of the National Hurricane Center (NHC) best track data for the North Atlantic and Eastern Pacific the Joint Typhoon Warning Center (JTWC) for other basins.

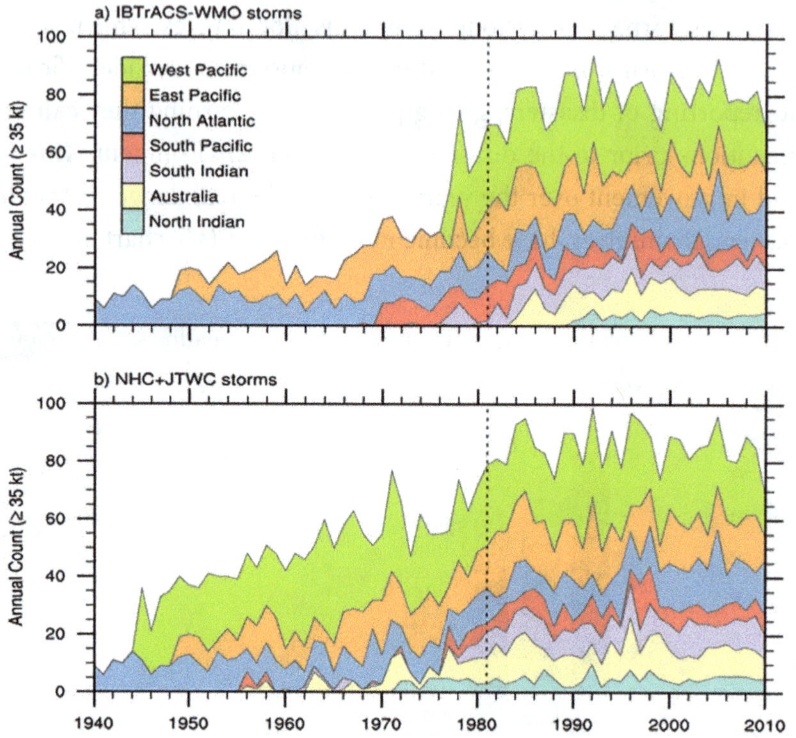

For the most part, increases in global storms have increased since 1940 in most of the regions of the world, West Pacific, East Pacific, North Pacific, moderately in the South Pacific, and the South Indian Ocean region. Only the Australian and North Indian regions have maintained a lower-level incident of disaster storm incidents.

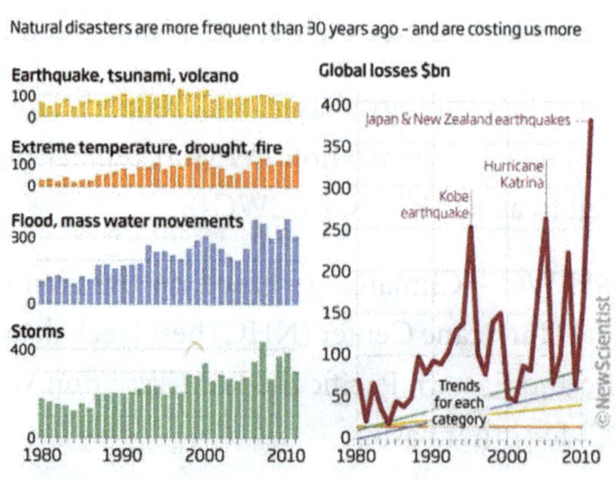

Impact on natural resources

Because fossil fuel emissions have caused changing weather conditions and ocean temperatures, this has effect on agriculture, fish, wildlife, water and energy. Crop yields affected by temperature and water stress as well as length of growing season has fallen by 10-25% and are less predictable as key regions shift from a warming to a cooler weather condition. As some agricultural pests die, due to temperature changes, other species spread more readily due to the dryness and windiness- requiring alternative pesticides or treatment regiments. Commercial fishermen that typically have rights to fish in specific areas will be ill equipped for the massive migration of their prey (Mgbemene, Nnaji, & Nwozor, 2016).

Impact on carrying capacity

Today, carrying capacity refers to the ability of the world, its natural resources, and its natural Eco-systems, including social, economic, and cultural systems to support the finite number of people on the planet, is being challenged around the world. According to the International Energy Agency, global demand for oil will grow by 66% in the next 30 years, but it's unclear where the supply will come from. Clean water is similarly constrained in many areas around the world. Climate professionals point to technological innovation and adaptive behavior as a means for managing the global Eco-system. It has been technological progress that has increased the carrying capacity over time. Over centuries we have learned how to produce more food, energy and access more water. But will the potential of new technologies be enough when a crisis like the one outlined in the scenario below occurs?

> *"Abrupt climate change is likely to stretch carrying capacity well beyond its already volatile limits. And there's a natural tendency or need for carrying capacity to become realigned. As abrupt climate change lowers the world's carrying capacity, aggressive wars*

are likely to be fought over food, water and energy. Deaths from war as well as starvation and disease will decrease population size, which over time will re-balance and realign the carrying capacity" (Mgbemene, Nnaji, & Nwozor, 2016, p. 19; IEA, 2020).

Impact on world security

Violence and disruption stemming from the stresses created by abrupt changes in the climate pose a different type of threat to national security than exists today. Military conflict can occur out of the desperation caused by the need for energy, food and water which is different than past violent military conflict caused by ideological and political differences, religion or national honor differences. "*The shifting motivation for confrontation would alter which countries are most vulnerable and the existing warning signs for security threats*" (Mgbemene, Nnaji, & Nwozor, 2016, p. 19).

Impact of Climate Change on health

Climate change causes environmental conditions that that have a negative impact on human health and existence, such as (direct effect): diseases and deaths are a result of extreme weather events like heat, flooding, mud slides, storms and hurricanes. Indirect effects stem from changes in the Eco-system, such as: conditions that facilitate infectious diseases, changes in agricultural production and the lack of clean water availability. Climate change can also have indirect effects on health from social and economic conditions caused by drought, flooding, famine, epidemics and human migration, movement of refugees.

Extreme weather conditions normally associated with climate change, apart from causing physical discomfort can also affect the immune system as well as cause the proliferation of diseases and decrease the resistance to disease. Many infectious diseases such as those caused by viruses and bacteria and spread by mosquitoes and rodents are influenced by seasonality and changes in temperatures. Change in rainfall and humidity are environmental results that can impact humans. Based on history of evolution it may be possible that climate change may cause genetic changes and mutation, not only in pathogens but also in humans. There may be a genetic change that results in increased virulence and resistance to antibiotics. Pathogens have been known to adapt to antibiotics and become resistant to traditional treatments. Climate change increases the potential to create an environment conducive for disease outbreaks. As survival becomes more stressful the impact on mental health becomes a real possibility (Mgbemene, Nnaji, & Nwozor, 2016).

> Frederica Perera is a leading expert on health at Columbia University, in New York City. *She runs the Columbia Center for Children's Environmental Health. "She and her colleagues have been studying the health effects of pollution and stress. Their findings have shown, again and again, that both can produce measurable harm to children"* (Perera, 2016; Grossman, 2016).

The highest risk youth are those who live in poor neighborhoods and children of color. They face a higher risk of breathing polluted air. They do not always have enough food to eat and their unstable conditions can lead to stressful life experiences. Children under the age of five years old are only 10 percent of the world population however, they experience 40 percent of the environmental caused diseases, according to the World Health Organization (WHO) (Fleischer, 2014).

The burning and emission of fossil fuels into the environment by cars, power plants and factories usually claim the health of the most vulnerable in our society, who are children. A child's immune system serves to defend against infections, disease toxic chemicals and poisons has not fully developed. As a result, the body of a child is not fully able to protect them from the harmful effects of breathing pollutants in the air. The pollutants in the air when inhaled by a child can potentially affect their brain (Perera, 2016). Global warming contributes to the spread of infectious diseases. Several recent studies have linked hotter air temperatures to an increase in diseases of the lungs and heart. Global warming contributes to the spread of infectious diseases. In a study conducted by Kotcher, Maibach, and Choi (2019), among 1644 adults were most concerned about the neurological damage that air pollution can potentially cause on babies and children. This study gives substantial data that supports efforts to address the neurological effects of air pollution on babies and children.

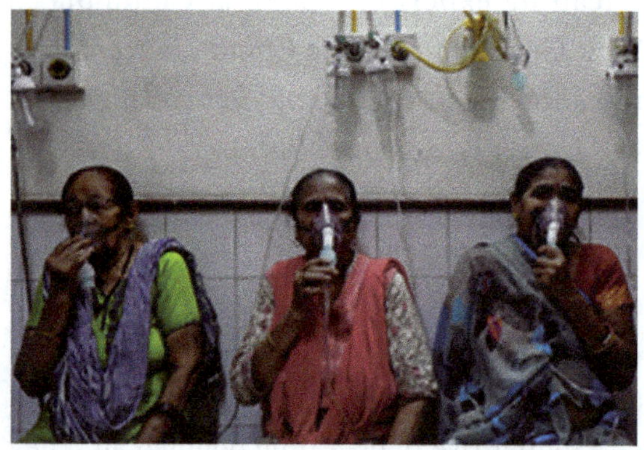

In New Delhi, India women are hospitalized and treated for respiratory problems caused by breathing the pollutants in the environment. People with chronic and contagious diseases are at higher risk.

A substandard diet, impure water and water shortage increases the risk of pandemics and sickness arising from environmental conditions (Chebreyesus, 2019). According to Chebreyesus (2019). Black carbon, methane, and nitrogen oxides are powerful drivers of global warming, and, along with other air pollutants such as carbon monoxide and ozone, they are responsible for over seven million deaths each year, about one in eight worldwide (WHO, 2020). Chebreyesus (2019) attributes man-made climate change over the last four decades has unleashed dozens of new infectious diseases, including Zika and Ebola viruses. New viruses such as the dengue infects 96 million people each year and causes 90,000 deaths. Yellow fever, Zika, West Nile, and other viruses that are spread by mosquitos are predicted to only get worse because of global warming producing an environment where mosquitos flourish.

One of the more critical impacts of global warming is the unsafe quality of drinking water in the world. Drinking water that contains unsafe levels of contaminants, can cause health effects, such as gastrointestinal illnesses, nervous system or reproductive effects, and chronic diseases such as cancer. Factors that influence the health effects and exposure of contaminated water are the concentration of contaminants in the water, an individual's susceptibility, the amount of water consumed, and the duration of exposure (EPA, 2022).

Common sources of drinking water contaminants include:

- **Industry and agriculture.** Organic solvents, petroleum products, and heavy metals from disposal sites or storage facilities can migrate into aquifers. Pesticides and fertilizers can be carried into lakes and streams by rainfall runoff or snowmelt; or can percolate into aquifers.
- **Human and animal waste.** Human wastes from sewage and septic systems can carry harmful microbes into drinking water sources, as can wastes from animal feedlots and wildlife. Major contaminants include Giardia, Cryptosporidium, and E. coli.
- **Treatment and distribution.** While treatment can remove many contaminants, it can also leave behind byproducts (such as trihalomethanes) that may themselves be harmful. Water can also become contaminated after it enters the distribution system, from a breach in the piping system or from corrosion of plumbing materials made from lead or copper.

- **Natural sources.** Some ground water is unsuitable for drinking because the local underground conditions include high levels of certain contaminants. For example, as ground water travels through rock and soil, it can pick up naturally occurring arsenic, other heavy metals, or radiosondes.

In additional to the problems climate change / global warming causes to physical health, evidence indicates that warmer temperatures can lead to mental health crises. According to the results of 53 studies, "researchers found significant increases in hospital and emergency visits due to mental health problems and in deaths due to mental health conditions during heatwaves. Specifically, mental health deaths increase by 2 percent every time the temperature increases by 1 degree Celsius" (Woods, 2022). Extreme heat is thought to contribute to mental health problems related to substance abuse and cognitive impairments, such as dementia. Deaths from preexisting psychiatric illnesses triple compared to other preexisting conditions during heatwaves. The risk of suicide also increases as the temperatures rise. Older individuals are at high risk of mental health problems during heatwaves, because their bodies are less capable of handling stress. People living in tropical and sub-tropical climate zones are also at higher risk.

A study, "published in February in *JAMA Psychiatry,* found an 8 percent increase in emergency department visits for mental health problems on the hottest days of the year compared to the coolest days. People are more likely to come to the hospital for self-harm, substance use, anxiety, and mood disorders as temperatures increase" (Amruta et Al., 2022).

Medical experts indicate there is "evidence that higher temperatures affect brain chemicals, specifically serotonin and dopamine levels which regulate mood, cognitive function, and our ability to perform complex tasks. Higher temperatures also lead to irritability and psychological distress, which are contributing factors to substance abuse and suicide. There is also clear evidence that exposure to high temperatures negatively impacts cognitive function, which likely explains the increased hospitalizations for dementia" (Mullins & White, 2019). In addition, extremely hot weather disrupts sleep and keeps people from sleeping. Sleeping problems are linked to nearly all mental health problems, and also lead to irritability, frustration, and poor mood. This is why higher overnight temperatures pose a serious threat to our mental health (Woods, 2022).

Effects of Global Warming on Violence and Aggression

Survival causes people to behave in unpredictable ways. When extreme weather conditions begin to threaten food supplies, water availability and life in general, people will migrate to other areas to survive. Whether they become refugees, interlopers, or vagrants will not matter. Movement and migrations of this type and cause will increase the potential for violent behavior, especially in already-vulnerable regions. Many scientists firmly believe that climate change will increases migration and conflict (Miles-Novelo & Anderson, 2019).

Countries that are in conflict and experiencing turmoil are especially vulnerable to violence and conflict. Countries with high population densities that experience environmental disruptions effecting loss of land due to floods, water contamination, crop shortages, and livestock depletion are susceptible to aggression and violence due to climate change. Economically advantaged countries will not escape the negative impact of climate change. Climate change may have an even more drastic impact on advantaged countries because they have no experience in surviving in disadvantaged conditions. Advantaged countries have experienced natural disasters but they have also been able to recover with the help of government subsidy. They have the hope of recovery. However, when hope is not an option then adaptation to living in disadvantage will not be easy. Changing their way of life, survival in dismal conditions with children, risk of exposure to disease and lack of food and water are situations Western countries are not prepared or conditioned. Risk of violent behavior increases. Crime data found that in America the disadvantaged neighborhoods of American cities, hot temperatures contributed to an increase in violent crime. Scientists and scholars speculate that the effect of climate change in regions, countries, or even neighborhoods will have a high likelihood of being breeding grounds for terrorist and gang activity. Possibly causing small-scale wars to break out across the globe (Miles-Novelo & Anderson, 2019)

CHAPTER VII

Denial of Climate Change

One perspective psychologists have of climate change denial is: the refusal to accept facts is to protect one from uncomfortable truths. This is considered to be a primitive defense mechanism. Anthropological climate change is a real phenomenon that is obviously apparent based on the occurrences in our environment. Many scientists are on record stating that, if not mitigated, will cause terrible suffering and destruction before the end of this century. A recent report from the United Nations-sponsored Intergovernmental Panel on Climate Change tells us we can still hope to avert some of the catastrophic consequences of climate change, but only if we "abandon coal and other fossil fuels in the next decade or two." Except for a small number of outliers, most scientists agree that we are rapidly approaching climate catastrophe (Gorman & Gorman, 2019).

A few uninformed politicians believe climate change predictions are exaggerated or even fabricated. More alarming is the lack of voice and complacency from the public. Perennial hurricanes in the Caribbean, the gulf coast and the east coast and West coast forests fires, the frequency of which indicates environmental disturbance. This should be enough for humans to take notice and give serious attention to the problem of climate change. Some people deny that climate change is a problem at all and don't admit that it represents a significant threat to civilization. One would think that evidence presented by climate scientists is right and that would be enough to engage corrective action. Despite meteorologists' reports telling their audience, the warmest temperature in recorded history is being experienced or reporting the occurrence of extreme weather conditions or polls that indicate the public believes climate change is real and an important issue, still little has changed in taking serious action. Even these facts don't motivate society to take serious action. It's pitiful how societies plant the seeds of their own destruction then when they get to the devastation point of no return, they frantically scuffle, engage behaviors that create chaos, then attack each other in ways that aggravate the destruction. The real pity is that the worse could have been avoided had society addressed the problems early on.

According to a research study at the Annenberg Public Policy Center (APPC), experiencing extreme weather is not enough to convince climate change skeptics that humans are damaging the environment. Political identity and exposure to partisan news influence people's position regarding their viewpoint on experiencing extreme weather conditions.

> *"Their research found that Republicans are less likely to report experiencing a polar vortex, while those exposed to liberal media are more likely to do so."*

Climate change has become a political issue that has diluted its importance as an issue critical to human survival (Lyons, Hasell and Stroud, 2018).

Hurricane Harvey's assault on Houston, Irma catastrophic destruction on Florida and the way Hurricane Katrina devastated the gulf region are considered typical of the extreme weather events that are likely to become more common as the planet warms.

> Victor, Obradovich and Amaya (2017), blame the cause of a relative unresponsive action on the hypothesis that *"Humans aren't well wired to act on complex statistical risks. We care a lot more about the tangible present than the distant future."*

Nevertheless, the public response to the findings of scientists is weak in demanding climate action by politicians. Behavioral scientists call this cognitive aspect of brain function, *"hyperbolic discounting."* This makes it difficult to conceive climate change as a threat, when the most significant danger is in the future, years ahead. Most voters know little about the details and facts of climate change, or the policy options relating to it. Instead, "voters' opinions about such things derive from heuristics such as political party affiliation and basic ideology." As stated, the majority of those who are defensively neglectful about climate change are either rich, wealthy, Republicans, protecting their financial interest or lower-class Republicans who deny or lack the cognitive mentality to understand rational logic. Poor lower-class White Republicans are so focus on supporting Republicans divisive and discriminating policies that they allow themselves to be blindly led, synonymous to the pied-piper leading children over the cliff. Voters do a poor job of "holding politicians accountable when the effects of political inaction are far removed from the policy failures that cause them" (Victor, Obradovich & Amaya, 2017).

Van Boven, Ehret and Sherman (2018) advance there are psychological barriers to bipartisan political and public support for climate policy. Political Scientist study the political polarization that fracture the collaborative effort for Republicans and democrats to agree on an environmental policy.

"If I proposed something that was literally word for word in the Republican Party platform, it would be immediately opposed by eighty to ninety percent of the Republican voters. And the reason is not that they've evaluated what I said. It's that I said it."

President Barack Obama (Remnick, 2016;
Van Boven, Ehret, & Sherman, 2018).

While Republicans are branded as climate control skeptics, results from national panel studies in 2014 and 2016 indicate that most Republicans believe in climate change, though not as strongly as Democrats. Political polarization over climate policy does not necessarily reflect that Democrats and Republicans disagree about the phenomenon of climate change but that Democrats and Republicans disagree with each other, both as citizens and policymakers. For example:

> *"In my first six years in Congress from 1993 to 1999, I had said that climate change was hooey. I hadn't looked into the science. All I knew was that Al Gore was for it, and therefore I was against it."*—Republican Representative Robert Inglis (2015) Van Boven, Ehret, & Sherman, 2018).

The statements above support the reason climate change must be taken out of the political realm. It's frightening to think that political differences could be the reason our world is destroyed, when humanity could have been saved with bipartisan and collaborative effort between elected officials. Consider the lack of bipartisan collaboration on the "corona-virus relief package," critical to the immediate needs of American citizens. The tragedy is that at a time when American citizens are suffering politicians act political. The needs of the American people don't weigh in their embattled relationship. This is an example of politicians not serving the people, but their own interest for retribution and revenge. It's deplorable (LeVine & Ferris, 2020). How many lives could have been saved if the corona-virus pandemic were ethically and competently administered? Compare the analogy to climate change. The relationship between Democrats and Republicans is so fractured that they are unable to collaborate on issues reflecting the public interest and their common interest. (Van Boven, Ehret & Sherman, 2018).

Most Republicans believed in climate change (70% and 63% in 2014 and 2016, respectively), along with most independents (78% and 70%, respectively) and most Democrats (93% and 89%, respectively). In the 2016 American National Election Study, a nationally representative survey of approximately 4,200 respondents, showed similar results. Most Democrats (90%), independents (84%), and Republicans (69%) agreed that global warming "has probably been happening" more than they agreed it "probably hasn't been happening." And the majority of Democrats, independents, and Republicans (89%, 85%, and 71%, respectively) agreed that, assuming it was happening, human activity was either equally or mostly responsible for global warming, in addition to natural causes (Van Boven, Ehret & Sherman, 2018).

The political disparity to act on climate change considering the reality of bipartisan belief is not so much reflective of the political party system as it is of the people who serve the political system. It is a condemnation of politicians, their ethics and a reflection of their lack of responsibility and duty to serve the people. It's political incompetence, plain and simple. This political disparity reinforces the importance of voters to make better voting choices that make politicians more accountable. Ordinary people believe that placing party over policy reflects poor citizenship. (Van Boven, Ehret & Sherman, 2018). Politicians don't seem to be able to resolve the fractured polarization of their stalemate regarding climate change, voters need to take initiative and vote for candidates who support climate change and who will act on that support.

Political Scientist who study and track voters and issues they support, attribute voter decision-making, tends to reflect *"retrospective voting"* habits on certain issues. Climate change and environmental issues fall into this category. The theory of *"retrospective voting"* basically states that voters base their decisions on their perceptions of the past performance of the parties and candidates in office. In other words, when things are going well, incumbents are supported. When things are going badly voters support, the challenging party (Campbell, Dettrey &Yin, 2010; Victor, Obradovich & Amaya, 2017). Voters evaluate the immediate visible results of a party or candidate. A most common focus of voters is the economy. Futuristic circumstances, such as the potential hazards of climate control, don't weigh into contemporary voting decisions. Voters look to the past in deciding how they will vote. Rarely do they look into the future unless they are confronted with an immediate undesired situation (Campbell, Dettrey &Yin, 2010).

The state of California's position on climate actions indicates an exception to the neglect by the Trump administration. However, climate change resolution does not rest with one state or one country. Global warming is a world-wide phenomenon that requires global intercontinental collaboration to resolve.

A majority of the American public report being concerned about climate change, but ranking their priorities, they rarely put climate policy in a median position on the list. The public does not overwhelmingly indicate that it is willing to spend what is needed to address the problem. Voters need to be better educated on the subject (Funk & Hefferon, 2019). What people believe does not always align with their behavior.

Thirty-six percent (36%) of Americans who are personally concerned about global warming / climate change, Republican or Democrat, are much more likely to see climate science as settled, between climate scientists. The most common pattern is that people's level of science knowledge, or of education, has either no significant effect or only a modest effect. Political party association and ideology are primary concerns in their viewpoint regarding climate change. This indicates climate change is more of a polit-

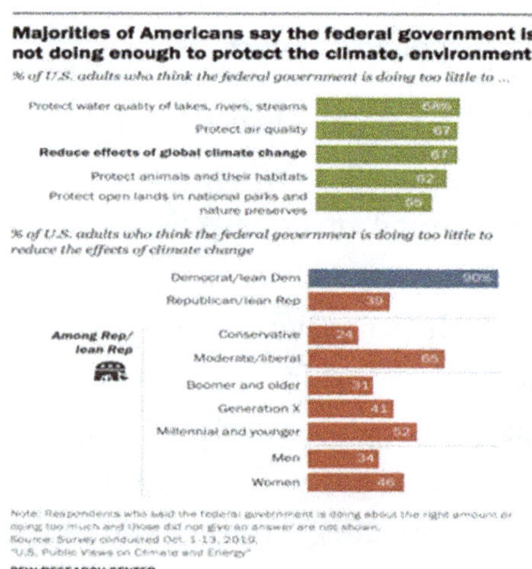

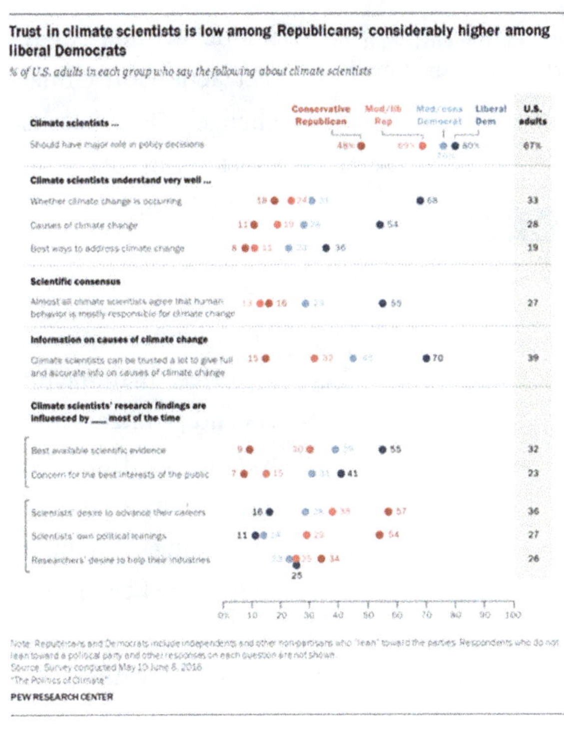

ical issue than a pulic safety concern (Funk & Hefferon, 2019).

The majority of Americans say that climate scientists should have a role in policy decisions about climate issues. Two-thirds (67%) of U.S. adults say climate scientists should have a major role and 23% say they should have a minor role. Just 9% say climate scientists should have no role in policy issues regarding global climate change. Political opinions on climate issues don't address the role humans play in addressing the issue, including their basic trust in the motivations that drive climate scientists to conduct their research (Funk & Hefferon, 2019).

Three-quarters of Americans (75%) say that they are particularly concerned with helping the environment as they go about their daily lives, while 24% say they are not particularly concerned. But only one-in-five (20%) of Americans say they try to live in ways that help the environment "all the time" (Funk & Hefferon, 2019).

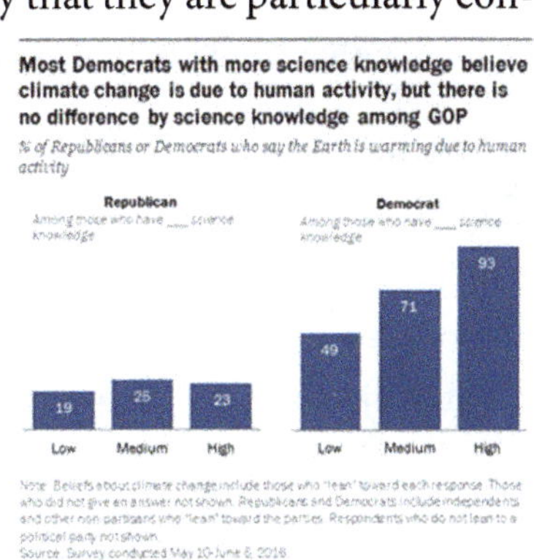

Realistically addressing one of the more dominate reasons people and

Most Americans report concern for the environment; one-in-five try to act on that concern all the time

% of U.S. adults who say that they are ____ about helping the environment as they go about their daily lives

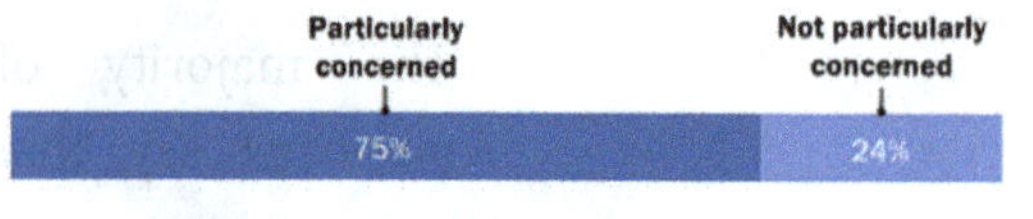

% of U.S. adults who say they make an effort to live in ways that help protect the environment...

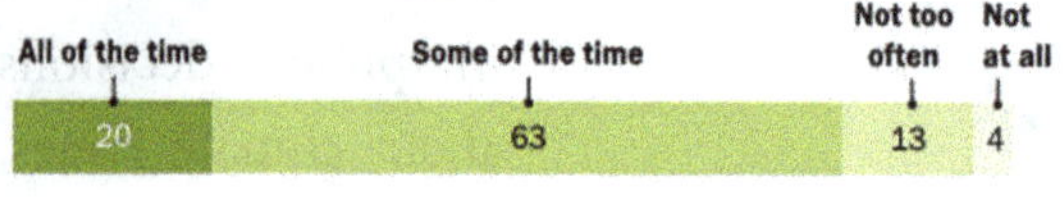

Note: Respondents who did not give an answer are not shown.
Source: Survey conducted May 10-June 6, 2016.
"The Politics of Climate"

PEW RESEARCH CENTER

politicians deny or reject action on climate change is because quick, deep cuts in fossil fuel emissions requires expensive costs the existing well-organized interest groups, factories, manufacturers and profiteers don't want to pay. The resulting benefit of seriously addressing climate change today, that will be executed internationally for beneficial results that will be realized in the distant future, is not appealing to contemporary leadership. Their value system is based on the "here and now." Failure to address emissions control, will cause society to deal with the politics of adaptation, to abandon vulnerable regions and subsidize the construction of various forms of protection, like sea walls to deal with rising sea levels and storm surges. Cities will need to develop defenses against hurricane, tornado and severe weather conditions (Victor, Obradovich, & Amaya, 2017). The human mentality is constructed to deal with reactionary responses rather than constructed aforethought to deal with a solution that solves the problem.

Another reason psychologist attribute to the refusal to accept the reality of climate change is what is referred to as *"motivated interference,"* which occurs when we hold a specific bias to ignore evidence (Gorman & Gorman, 2019). This can include a general unease with large government projects that are expensive and interfere with individual lives and livelihood. For example, people in the oil industry, whose livelihood is dependent on fossil fuel emission, resent government taking money out of individuals' pockets in the form of public spending on carbon control programs (Gorman & Gorman, 2019; Mortillaro, 2018).

Simply providing people with the facts about climate does not necessarily change people's minds. The science that empirically proves the earth is warming is very technical and difficult for most individuals to understand. Gorman and Gorman (2019) argue, "Humans aren't well wired to act on complex statistical risks."

> *"In addition to motivated interference, there is also a powerful psychological component to this blindness to scientific reality: denial. A lot has been written about climate change denial and there are clearly many explanations for it. For one thing, an enormous amount of money is being spent encouraging us to ignore climate change. Corporations, especially the fossil fuel industry, have spent huge sums attempting to obfuscate the reality of climate change. We are constantly told by them that "more data are needed" because "climate scientists don't agree." While no scientist would ever disagree with a call for more research—that line is, after all, found near the end of almost every scientific paper ever written—it just is not true that scientists don't agree that climate change is real. To some extent, then, we are the victims of a well-funded and sophisticated misinformation campaign that attempts to keep us in the dark about climate change" (Gorman & Gorman, 2019).*

Even when the evidence about climate change is presented in a clear concise manner, including basic terms with lots of compelling graphics, many people either don't believe it or shrug it off. Therefore, the problem of climate change denial is not simply a matter of an information gap, it's too much too be comprehended and too complex for most people to fully understand.

Among the array of reasons that people deny the problem of climate change, is that its enormity becomes mind boggling. What the climate scientists are telling us is that if we don't stop burning fossil fuels humanity faces extinction. It's not comparable to a bank account that is overdrawn, a car that was stolen or the thought that the roof caved in on your house. Potential calamity can be resolved when seen in a perspective of manageable remedy that can be handled in reasonable short order. A problem that requires this magnitude of solution and that will require decades of life changing effort to resolve, does not motivate people to act. It's easier to deny the reality of climate change (Gorman & Gorman, 2019). What happens to people when it's too late to reverse tragedy?

Denial on this scale can have monumental repercussion. The phenomenon of climate change affects millions, worldwide. This is not just an individual or group psychotherapy situation it is internationally widespread. The effort must be a collaboration of industrial nations because fossil fuel emission is an international problem. Some people are just gullible and naively convincible. They support political candidates and become blindly supportive of every issue those candidates support. Political candidates take money from special interest groups and blindly support every issue the special interest stand for, even if it is not in the best interest of America. Ultimately, only a large-scale collaborative international political effort can save civilization. Nations must come together as they did in forging the Paris Climate Agreement and agree to enforce what will have to be very extensive and often highly inconvenient changes in our sources of energy and food. Unfortunately, the Republican administration of Donald Trump canceled America's participation and involvement in the Paris Climate Agreement. Replacing oil and gas with sustainable energy and switching to plant-based diets will be difficult and even painful for some, but the alternative, continuing to ignore that climate change is already affecting us and will ultimately be catastrophic, will be much worse and have catastrophic consequences for humanity.

Chapter VIII

The Effect of Human Psychology on Climate Change Response And The Effect of Climate Change Response on Human Psychology

The essence of psychology is to understand how people think and how their thinking correlates and aligns with their behavior. Everyone or at least a significant number of people in the world and most leaders of world nations, are aware there is a climate change problem. They are aware of the high potential for climate change to have a devastating and destructive effect on our world. Their thinking recognizes that climate change and global warming is a problem with significant potential risk to humanity. Why then do some government officials and business people fail to act, with the acknowledgement, that climate change is a serious environmental issue? The Intergovernmental Panel on Climate Change (IPCC) issued a press release in February 2022 stating that "Human-induced climate change is causing dangerous and widespread disruption in nature and affecting the lives of billions of people around the world, despite efforts to reduce the risks." The same report also indicated the window for taking corrective action is "narrowing" and immediate action needs to be taken (IPCC, 2022). Despite data and evidence indicating global warming warrants serious concern, it is a dilemma that the United States. and the world responds casually. One can speculate that business people dismiss the seriousness of global warming because it could have a negative impact on their business bottom line, if the problems of climate change were seriously addressed. Ex-President Donald Trump has denied climate change and global warming as an issue important enough to give serious attention. That he pulled the United States out of the Paris Agreement is telling. If one asks why, at least three reasons come to mind. First, one can hypothesize Trump is on a mission to undue all the Obama era accomplishments that he can. Many people believe Trump is a racist and obsessed with jealousy and envy regarding President Obama because President Obama is the first Black President of the United States of America. A second reason, and not to discount the first, is that Trump is just too plain stupid to understand the seriousness of the issue. A third reason is also a possibility. Trump is so focused on money and business economics; he has a narrow-minded and self-serving perspective focused on providing support to the wealthy and the business community. He simply ignores issues that do not serve the purpose of advancing the success of his business interests. His thinking ignores and at the least, diminishes, the issue of global warming. His position on climate change indicates his only concerns are immediate without any prospect for the future. His concern for the common person, the American citizen, with limited wealth is low priority. Longstanding views that Trump is unfit to be President seem to be valid when evaluated in perspective with his views on global warming and climate change. Even though Donald Trump is no longer president of the United States. It is relevant to express his views and position because there are still a significant number of Americans who follow him. Joe Biden is the current United States President as of the 2020 election and he has made every effort to restore America back on track to address climate change issues. Nevertheless, the obstructionist who follow the "Trumpian" philosophies are problems to the issue of addressing climate change and global warming.

The fact of the matter is that climate change is already killing people. Then why don't we address the issue as a pandemic? King, (2019) is saying that people are conditioned to think in terms of their immediate situation and give less priority to preparation for the future.

> *King, (2019) suggest "cognitive biases that ensured our initial survival now make it difficult to address long-term challenges that threaten our existence, like climate change."*

A report from international climate experts tells us that we are likely to reach 1.5C of average global warming in approximately 11 years. (IPCC, 2018). At that point we can expect *"increased risks to health, livelihoods, food security, water supply, human security and economic growth"*. These experts also found that temperature rise has already altered human and natural systems in profound ways, resulting in more extreme weather, the melting polar ice caps, sea level rise, droughts, floods and biodiversity loss. However, even this information has not effectively caused humans to change their behaviors significantly enough to alter climate change. And a big part of the reason is our own evolution. The same behaviors that once helped us survive are, today, working against us (IPCC, 2018).

Cognitive biases that initially supported human development and growth, now hinder the long-term challenges that threaten human existence, specifically regarding climate change. Climate change is already killing us. Psychologists have identified numerous cognitive biases commonly shared. King (2019) succinctly explains a few specific reasons people lack the will to address climate change.

- ***"Hyperbolic discounting.*** This is our perception that the present is more important than the future. Throughout most of our evolution it was more advantageous to focus on what might kill us or eat us now, not later. This bias now impedes our ability to take action to address more distant-feeling, slower and complex challenges."

- ***"Our lack of concern for future generations.*** Evolutionary theory suggests that we care most about just a few generations of family members: our great-grandparents to great-grandchildren. While we may understand what needs to be done to address climate change, it's hard for us to see how the sacrifices required for generations existing beyond this short time span are worth it."

- ***"The bystander effects.*** We tend to believe that someone else will deal with a crisis. This developed for good reason: if a threatening wild animal is lurking at the edge of our hunter-gatherer group, it's a waste of effort for every single member to spring into action, not to mention could needlessly put more people into danger. In smaller groups, it was usually pretty clearly delineated who would step up for which threats, so this worked. Today, however, this leads us to assume (often wrongly) that our leaders must be doing something about the crisis of climate change and the larger the group, the stronger this bias becomes."

- ***"The sunk-cost fallacy***. We are biased towards staying the course even in the face of negative outcomes. The more we've invested time, energy or resources into that course, the more likely we are to stick with it – even if it no longer seems optimal. This helps explain, for example, our continued reliance on fossil fuels as a primary source of energy in the face of decades of evidence that we both can and should transition to clean energy and a carbon neutral future."

The IPCC has encouraged psychologists to become part of the integrated scientific effort to support the achievement of climate change targets such as keeping within 1.5°C or 2°C of global warming (Nielsen et al. (2020). A benefit of involving psychologist in the effort to curtail climate change is because to effectively address the problem requires human mental and behavior change. The challenge for psychologist in addressing climate change is that it is a society problem that involves massive populations of people. Any change in cognitive thinking and the subsequent behavioral change is far more complicated than one-on-one office consultations. How do you change the thinking and behavior of an entire society, entire societies? Addressing climate change requires a massive education effort. People need to learn new habits and new ways of behavior.

In order to effectively grasp and behave with serious intent, regarding climate change, people need to have a transcendental ability, or there needs to be a catalyst that induces people to be able to transcend their immediate situation and envision the future in a world where they have become slaves to the environment. A world where humanity no longer controls their daily lives but are living in a destitute state of survival humans, themselves, created. A tragedy is that future generations will be victims of a devastated world condition they had no hand in creating. It's ironic that politicians are concerned with a hard-core focus to get the nation's budget under control. They rationalize their failed conservative views to control government spending on the premise of easing the debt on future generations. Yet, they fail to protect future generations from a potentially greater devastation. The conflict in the theoretical conservative philosophy to give future generations a debt free future does not align with common sense. Even if conservatives were successful to secure a debt free future for tomorrow's generation, the future will be a bleak wasteland where money will be the least valued commodity. Food, safe water, good housing, breathable air will be what people will demand. This scenario speaks to the lack of foresight and lack of intelligence by climate change obstructionist. People who obstruct efforts to control global warming should have no voice or influence in the lives of people. Unfortunately, their efforts to control the national debt have proven to be futile and a failure. Yet the very same politicians who have failed drastically to control the national debt make it obvious they have no concept of how to govern a city, state or country. The truth is these people don't care about humanity. They only care about their own selfish short-sighted greed. The voting public must take a large share of the responsibility for allowing these obstructionists to be in office.

Psychologist can help people to construct and develop the mindset to address climate change and help people adjust to the required changes in behavior that are necessary. Climate Change requires behavior change. Solving the global climate change crisis is going to rely on, in one way or another, changing human behavior (Williamson et al., 2018). Climate change is at a critical stage. Reversing the warming process is not a "quick fix." The problem did not develop overnight and won't be fixed overnight. Conditioning people to develop a cognitive mental construct regarding climate change in terms of a long-term perspective will be challenging. Psychology can contribute to helping people develop a mindset that will enable them to better adjust to the necessary and required behavior changes. An inhibiting characteristic among Americans is an underdeveloped sense of interdependence between individuals and groups. Now more so, or as much as any other time in history, Americans need to relate as Americans, not as Whites, African Americans, Italians, Germans, Polish, Hispanic or Asian but as Americans. If climate change is not addressed in a timely manner, scientist forecast a "dooms day scenario" (McKENZIE, 2019).

When one literally thinks about it, the reason people fail to address climate change is categorized in the same index as the reasons people don't save for old age because they are not future sacrifice oriented. We live in a now society. Enjoy life now and presume the future will take care of itself. Consider people who sell drugs. Most drug dealers believe they will never be caught, and they will make immediate quick money. The future will be guaranteed by the quick money they will make selling drugs. However, the truth of the matter is most always get caught and have their lives ruined. Those who recover from disdain go through tremendous recovery and rehabilitation to get back on a positive track.

Some who use drugs become addicted to the immediate gratification, many eventually overdose. People use sex to satisfy immediate gratifications without intention of marriage and having a family. If their parenthood is unplanned then it causes a disrupting life experience that requires greater sacrifice. People are generally oriented to pursue experiences of short-term gratification, especially when money is involved. The oil companies, companies responsible for the burning of carbon fuels in the manufacturing process, makers of plastic bags, and companies responsible for destroying forest trees that mitigate climate change, all risk the loss of millions / billions of dollars. If the world collaborated to enact climate change policies and follow lifestyle behavior changes adherent to clean energy policies, these companies would become obsolete and go out of business. In the immediate need to survive polluters lobby politicians with bribes, campaign contributions and special favors, inducing politicians to rival the cause of pollutant benefactors. This involves supporting legislation that maintains the survival of pollutant enterprises and suppressing legislation that supports clean energy. The polluters and politicians are tied irreverently to the short-term immediate gains of polluting the earth at the expense of the future of our children. Short-term inconveniences of the individual citizens to adhere to climate and clean energy behaviors will be challenging and much more inconvenient when compared to the comfortable ways of life already being experienced when natural disasters strike.

The most productive role psychologist can fill in mitigating climate change is to develop methodologies and programs that motivate voters to support climate change advocates for public office, locally, state-wide and nationally. Another role that demands the attention of psychology is to help citizens construct cognitive thinking compatible with behavior that supports changes necessary to mitigate climate change. The behavior changes necessary for individuals to adapt to a clean environment lifestyle will require mental and behavioral alignment. Psychology can play a crucial role in that area. Consideration should be given to develop climate change / global warming education in the schools, beginning at the elementary school level.

One way to learn how a person or group thinks about climate change is to understand their Environmental Values. One aspect of environmental empirical research is a process of exploring the relationship between cognitive values and environmental behavior. In the context of environmental behavior three types of values are important: **egoistic** (a focus on the self), **altruistic** (a focus on the welfare of others), and **biospheric** (a focus on the welfare of the environment). Biospheric refers to the part of the earth and its atmosphere in which living organisms exist or that is capable of supporting life. The living organisms and their environment composing the biosphere. Thus, an individual might be motivated to carpool because it is cheap (egoistic) or because carpooling has less emissions of fossil fuel with a lower impact on the health of other people (altruistic), or because carpooling has less fossil fuel emissions and less impact on the environment (biospheric) (Newell et al., 2017).

Studies indicate that egoistic values tend to be negatively related to pro-environmental attitudes, intentions, and behaviors. Altruistic and particularly biospheric values are strongly positively related to pro-environmental attitudes (De Groot & Steg, 2010). In the study by De Groot and Steg (2010) participants were asked to indicate their intentions to carpool and donate to environmental charities. Participants' environmental values were assessed along with measures of their self-determined motivations. Self-determined motivational types perform pro-environmental acts more frequently and engage in activities perceived to be more difficult. Values were more predictable for pro-environmental intentions than were self-determined motivational types, but values and motivational types were related to environmental concerns in several ways.

De Groot and Steg (2010) found that participants who strongly endorsed statements such as "I enjoy contributing to the environment," an example of intrinsic motivation scored high on measures of altruistic and particularly biospheric value orientations (i.e., including things like preventing pollution and respecting the earth as highly important). These findings led De Groot and Steg (2010) to recommend that environmental influencers should attempt to promote pro-environmental decisions based on appealing to people's biospheric values and intrinsic motivations. However, this strategy is less effective and less practical for group interventions. A problem with climate change strategies is to find ways to motivate participation by the mass populous (Newell et al., 2017). Understanding individual's environmental and cultural worldviews are a way to determine the psychological viewpoint of their environmental decisions. Exploring and understanding the worldview held by a person on the environmental, captures a person's general beliefs about the relationship between humans and their environment values (Newell et al., 2017)

A measure used to evaluate worldviews is the new environmental paradigm (NEP). The NEP includes questions about the beliefs that humanity can upset the balance of nature, that humanity does not have the right to rule over nature, and that there are limits to the growth of human societies. The NEP has been used to predict pro-environmental behaviors, intentions, and policy acceptance, but it does not appear to be as powerful a predictor as values, especially biospheric values (Newell et al., 2017).

Cultural Cognition Theory (CCT) merges traditional cultural theory regarding social relations with theories of how people form risk perceptions. This merging has led to the development of a concept by which a person's worldview is measured. Background data such as gender, race, and class are collected. Attitudes about social variables are assessed measuring CCT using statements such as, "*We need to dra-*

matically reduce inequalities between the rich and the poor, whites and people of color, men and women." Alignment with a society in which people should determine their own well-being without governmental assistance is measured, including information gathered by asking questions such as *"Should the government do more to advance society's goals? Should the government limit the freedom and choices of individuals to advance society's goals?"* The responses on these two questions defines individuals as hierarchical individualists or egalitarian communitarians (Newell et al., 2017). Hierarchical individualists or egalitarian communitarians classification of people's worldviews according to this CCT structure is related to the acceptance of scientific consensus views on climate change and disposal of nuclear waste as well as non-environmental issues, including gun control and the human papilloma virus (HPV) vaccine, according to these researchers (Newell et al., 2017). Hierarchical individualists are those who generally oppose government intervention and restrictions on industry and social issues. Egalitarian communicants normally support government intervention and restrictions on industry and social issues.

Evaluating CCT against the science comprehension thesis in an examination of attitudes toward climate change. According to the authors, the science comprehension thesis predicts that individuals who are more scientifically literate should be more concerned about the risk posed by climate change to human health, safety, and prosperity. However, ironically, according to a study by Kahan et al. (2011), there is no evidence that positively correlates scientific literacy to belief that climate change poses a risk to human health, safety, and prosperity. Data was collected from over 1,500 US citizens, which found evidence that cultural worldviews, not scientific literacy, influenced belief in climate-risk perceptions Kahan et al. (2011). The CCT study concluded that potentially it is likely for different cultural biases to coexist simultaneously in varying degrees within individuals. When communicating environmental information, one needs to consider the worldviews of the individual or group receiving the information (Newell et al., 2017).

Morality and political orientation often are factors that influence opinion about climate change. Environmental values correlate with worldviews, worldviews, in turn, correlates with political orientation. Hierarchical individualists tend to be more aligned with conservative political agendas, whereas egalitarian communitarians are more likely to share the views of political liberals (Newell et al., 2017). Stereotypically, Republicans align with the profile of hierarchical individualists and Democrats align with egalitarian communicants. In addition to their polarized positions on climate change, they also tend to have opposed viewpoints on civil rights, abortion, income inequality, government intervention in matters of equal opportunity, police brutality and most often Republicans are branded as racist.

Feinberg and Willer (2013) advance a hypothesis that the basis for this polarization of environmental views along political lines stems from differing perceptions of whether environmental concern is a moral issue. They suggest that liberals tend to resonate with "harm-and-care" based moral arguments, which appear to have media appeal when presenting environmental issues. Feinberg and Willer (2013) advanced that liberals exhibited more positive pro-environmental attitudes than did conservatives when confronted with climate change presented in terms of the harm humans are causing the environment and humanity, obligating the need for us to care and protect our world. However, when the message described the extent to which the environment had become polluted and emphasized the need to purify it, conservatives and liberals showed equally positive pro-environmental attitudes. These results suggest that political polarization of environmental views can be overcome, or at least reduced, by direct appeals to the moral nature of individuals (Newell et al., 2017).

The bottom line to mitigating and resolving the threat of climate change lies in behavior change in the lifestyles of all individuals all over the world. That is a humongous undertaking. How does the threat of climate change / global warming translate into individuals changing their behavior to offset disaster and even the potential threat to the end of human existence as we know it. The general held practical theory on behavior is that people don't change unless they are confronted with an immediate threat to their health, their life, their livelihood or other phenomena that is important to them. When threats are in the distant future people to not display or express the same urgency to change their behavior (CDC).

While King, (2019) is saying that people are conditioned to think in terms of their immediate situation and give less priority to preparation for the future, it is nearing the point where reliance on the freewill of people will not solve the problem. There needs to be government intervention that, unfortunately, regulates human behavior in relationship to environmental activity, if people persist in not giving serious attention to the problem of the earth's changing climate and global warming. The alternative to government regulation, as a reluctant alternative, is devastation to the earth, health crisis to people, and the loss of human life. These alternatives make government regulation a savior, in this case. It is unfortunate that humans need to be saved from themselves, but in this case, that is the case. Nevertheless, people will need to adapt to behavior change. Most of the human behaviors that contribute to global warming and climate change have been part of the human lifestyle and behavior patterns for centuries. Therefore, the issue becomes how do we break old habits? The real irony and surprising aspect of human nature is that even in the midst of a health crisis epidemic humans still rebel against taking safety precautions. Evidence of this is the movement on the part of multitudes of people to unmask, to overlook social distancing, and to ignore the dangers of interacting in large crowds before the Covid-19 pandemic was and is under control. Frankly, as of the first quarter of 2023, the Covid-19 pandemic is still not under control but people are staunch about being unmasked and interacting without regard for the fact Covid is still an active threat in the world and doctors predict it is here to stay, much like the flu.

For the most part, the problem is not caused by the many people who are dedicated to protecting the environment and acting in accord with environmental protection. The problem is with the people who operate in rebellious fantasy and do not acknowledge the seriousness of the problem. These people, just as those who disregard Covid-19 protections, are as much a treat as the Covid pandemic, the earth's changing climate, and the warming of the globe. So the question becomes how to handle the dissidents? Government regulations are associated with disciplinary actions, however, the preference is to align the thinking and behavior of dissidents with the concern for the preservation of earth. In other words, how do we change human behavior in the midst of no immediate threat, to avoid an immediate threat?

Ideally, the mitigation of climate change / global warming problem requires a combination of government intervention and consumers behavior change. Research indicates that 90% - 98% of people have awareness of the climate change / global warming problem. Over 80% of people express concern about climate change (Semenza et Al., 2008).

Research found that barriers such as skepticism, distrust, fatalism, and lack of knowledge, at the individual level, contribute to restrictive behavior change.

A number of cognitive and behavioral stages explains individual barriers to change, in addition to the structural barriers (individuals who make their livelihood in fossil fuel areas), other categories of reservations are:

- "**Pre-contemplation** (does not believe in climate change or the usefulness of behavior change): *I don't have control over* *something like global warming, that's not something a person can change"* (Semenza et Al., 2008).

- "**Contemplation** (believes in climate change but no action considered): *I can't move unless I'm in an automobile, and I still have to farm with tractors. Other than that, I can't change what I'm doing. I just can't change my behavior. I'm not going back to farming with horses"* (Semenza et Al., 2008).
- "**Preparation** (steps considered but no behavior change yet): *I know what things I could do to change, but because I live where I live and I work where I work, I cannot make the change, specifically, driving to work"* (Semenza et Al., 2008).
- "**Action** (minimal individual attempts but sees limitations and difficulties): *I don't tolerate the heat very well. I use the fans for as long I can stand it, and then I turn the air conditioner on"* (Semenza et Al., 2008).
- "**Maintenance** (maintains several actions): *I feel that I am doing what I can right now and can't do any more. I am happy with what I do now"* (Semenza et Al., 2008).
- "**Termination** (maintains low-carbon lifestyle and identifies with ideology): *I was already concerned about the environment and doing what I could to not pollute. We were already aware and conscientious about our actions"* (Semenza et Al., 2008).

People who live in areas where their livelihood is dependent on fossil fuel production were less willing to behave in "environmentally friendly" ways. Areas such as Texas (Oil country and West Virginia, Indiana (Coal country) were softer on behavior change than states in Northwest America. Individual behavior change will be easier to facilitate under favorable contextual conditions, and there would be less resistance if there were accompanying incentives from industry, commerce, and government. To make voluntary behavior change, incentives such as easy accessibility to clean energy options and economic incentives would be more likely to induce behavior change (Semenza et Al., 2008).

Another obstacle is the portrayal of climate change issues in the media. Media messages should be educational and reliant on the facts. However, too much focus is on the controversy. Media messages should focus on the facts that increase awareness about the causes and consequences of climate change. Many journalists operate on the opinion that there is confusion and disagreement within the scientific community. Media advocacy should be evidence-based, with unequivocal messages tailored directly to these segments of the population. Instead of stirring the pot of controversy, media needs to encourage solutions and eliminate the confusion they cause. Confusion and disagreement scenarios make for great press among journalist. They seem to be more willing to keep the conflict ongoing that to take a responsible position in favor of mitigating climate change / global warming.

Behaviors regarding climate change are variable. There are a significant number of people take that are not incentivized to change their behaviors. In spite of the consequences of neglectful behavior on the overall society, collective actions can be encouraged by economic incentives. While political psychology perspectives and social discourse do play a role in communicating the relevance of climate change / global warming issues and sensitize individuals to the problems but economics will motive them to change much more than rhetoric. One cannot blame this disposition because the same economic drive that cause companies to create the climate change problem is common for individuals to contribute to mitigating the climate change problem. Frankly, the companies that caused the problem should pay for alleviating the problem, without compromise. The problem is that serious.

Behavior change reluctance is not surprising from the perspective of economic theory, because "at the heart of most collective-action problems is the tendency for individuals to ignore the social costs and benefits of their decisions." Gun manufacturers continue to flood the society with guns regardless of how many people use guns to kill each other and children. A political behavior or political psychology perspective does contribute to the fact that people have a sympathetic awareness, influenced by the social and political theories and rhetoric. "If there is a real link between individual behavior and collective outcomes, it is possible for public discourse to increase or decrease the salience of that link." Economic incentives will only give motivation to latent, unmotivated or even resistant individuals to change their behavior (Lubell, Zahran, Vedlitz, 2007). Realistically, that is the American way. However, the entities with primary responsibility for addressing climate change problems (the government, corporations, manufacturing) are money motivated to create the problems. However, when it comes to resolving the problem they created, their motivation to provide the same economic incentives motivating them is reluctant.

A significant problem with behavior change among the masses is that scientists construct the information on climate change / global warming in language the masses do not understand and do not relate to (Nerlich, Koteyko & Brown, 2010). Climate change information should be delivered in terminology and language the masses can understand. Getting people to change their behavior in the midst of what appears to be normal environmental conditions for the most part is not easy. Natural disasters are increasing, cost of disasters is monumental and the death tolls are rising but for people who hit by tragedy, it is easy for them to behave as they always have. This is a reason that government regulations and rewards need to be enacted as motivational incentives. Better understanding of climate change / global warming issues will give the voting public a better perspective in evaluating politicians platforms and making decisions regarding their voting choices.

Truelove and Parks (2012) conducted a study and found that while individuals, especially adolescents and young adults are aware that the emission from cars causes pollution and is a major contributor to climate change / global warming. However, they did not express knowledge of the significance in behaviors such as recycling, adjusting the thermostat, the efficacy in not littering, and not eating meat in reducing global warming. As widespread as the issue of global warming is discussed in the society, the micro details get less focus than the macro details. Truelove and Parks (2012) suggest the micro details need coverage in climate change / global warming communication campaigns and behavior change strategies when informing the public. However, when discussing the micro aspects of climate change, it needs to be conveyed in language the non-scientific individual can understand.

The theories of behaviorism in psychology surmises that complex behaviors result from the basic elements of stimulus and response. In other words, behaviorism advances that an individual develops all aspects of behavior through experiences related to the connection between environmental stimuli and responses to those stimuli. Those responses may include cognitive elements, unobservable mental processes and choice in mediating the behavior of the individual (Heimlich & Ardoin, 2008).

Relative to the environment and global warming / climate change, if this is postulated, then why do countries and people lack the aggression to address climate change? The many occurrences and exorbitant cost of natural disasters occurring through the world should be significant warning of what is to come. It should also be adequate proof for people to change their behavior and address climate change / global warming seriously. However, this is not happening. Governmental bodies are addressing climate change strategically in terms of economic preservation and lifestyle preservation as compromise to the fact that natural disaster to humanity is nearby. Past and present experiences with the consistent flux of natural disasters is not sufficient motivation for people to sacrifice and address climate change aggressively. The receipt of benefits causes behavior change in today's society. People have a strong attachment to material possessions and are reluctant to sacrifice until they are forced to do so. It is like religion. People will profess to be religious but do not become expressively religious until they are about to die.

Noted psychologists Ivan Pavlov and B. F. Skinner are known for their work in the area of classical conditioning. Both Pavlov and Skinner theorized that changes in behavior are the result of an individual's response to events (stimuli) that occur in the environment.

> *"When teaching people about recycling, for example, some environmental educators might encourage them to recycle and teach them to use a recycling bin to separate plastic, paper and glass. The eventual desired outcome for that approach to teaching behavior is that the individual will recycle used items; however, this desired behavior risks breaking down if the person becomes reliant on the bin rather than committed to the recycling behavior. If that person is in a situation without the recycling bin, he or she may decide to throw recyclable items in the trash because the first-order behavior (sorting items into bins) is absent"* (Heimlich & Ardoin, 2008).

However, the most effective means of inducing behavior change is immediate reinforcement. In the context of contemporary society, the most responsive means to induce behavior change is economic stimulus. In addition. Often environmental educators focus on the behavioral outcomes rather than the steps required to reach those outcomes.

Most environmental activities are made up of several discernible behaviors' (recycling, car-pooling, do not litter, temperature control, dietary changes to include environmentally friendly foods and use of environmentally friendly products). Success is easier to achieve when individuals learn environmentally friendly behaviors early in life, rather than expect them to change traditions without beneficial incentive. Without an immediate threat, people need incentives to motivate behavior change. An immediate threat requires no outside incentives because people are personally motivated internally. In the case of climate change / global warming, it may be too late when personal motivation "kicks in."

Chapter IX

Money and Climate Change

It can be argued that the biggest and most problematic problem with human beings is that the leaders of countries and decision-makers who are responsible for the lives of others, value money and wealth more than human life. That is, they have more value for money and wealth and less value for the life of others.

The IMF's Fiscal Monitor group found that a global tax of $75 per ton by the year 2030 could limit the planet's warming to 2 degrees Celsius (3.6 degrees Fahrenheit), or roughly double what it is now. That would increase the price of fossil-fuel-based energy, especially from the burning of coal but the economic disruption would be offset by passing the cost increase to the consumers. In the United States, a $75 tax would cut emissions by almost 30 percent but would cause on average a fifty-three percent (53%) increase in electricity costs and a twenty percent (20%) rise for gasoline at projected 2030 prices, according to the analysis of the IMF's Fiscal Monitor found (Mooney & Freedman, 2019). Taxing companies that are responsible for producing greenhouse gas emissions is an idea economist have tossed around for a long time. However, politicians have obstructed taxing manufactures of fossil fuels. Nobel Prize-winning Yale economist William D. Nordhaus has advanced that a carbon tax of $300 per ton or even higher might be required to control carbon emissions (Mooney & Freedman, 2019). The problem is that capitalist also focus on money as the solution to their problems and the catalyst to their opportunity. They are brain dead outside of the realm of money. The Raising taxes, only raises cost to the consumer and will not solve the problem. The solution to the problem is to replace "dirty" energy with "clean" energy, to eliminate global dependence on fossil fuels. The only solution to the problem is avoided by focus on taxes and increasing the cost of fossil fuels. Besides, increasing the cost will also mean that people who cannot afford higher cost will suffer.

The International Money Fund (IMF) suggests that "*The world needs a massive carbon tax in just 10 years to limit climate change.*" This thinking shows the mentality of people who are in the money business vs people who want to see real solutions. Taxing the companies who pollute does not solve the problem. Where does the tax money go? The tax money goes to people who want more money, and the tax increase only gets passed on to the consumer raising consumer cost. The solution is to eliminate the pollution, not to charge more money for polluting. Money solution advocates have a narrow vision. Their mentality is limited. They think money is the solution to all problems. How much will it cost to buy your way out of dying? The consumer will wind up paying more money in health care and still suffer the effects of a polluted environment. The only way increased taxes might cause a positive change would be to increase taxes and mandate prices do not increase. Impacting profits is the only way to give pollution companies concern and reason to change their behavior. But even that is no comparison to eliminating fossil fuels altogether.

To reach the goals of the 2015 Paris climate accord (Appendix II), fossil fuel emissions will need to be cut in half by 2030 and eliminated all together by 2050 (McKibben, 2019). The elimination of fossil fuels will put the billion-dollar oil, gas and coal companies out of business. Automobile companies will need to change from gas cars to electric cars. Home heating and energy systems will need to change to operate with more energy efficiency. The multi-billion-dollar Koch company will be crippled and suffer great losses. Preservation of the Koch Industries Empire is likely a reason they are so politically active, in terms of funding candidates for political office. Some of the largest corporations in the world are oil companies. These billion-dollar companies are publicly traded on the stock exchanges around the world. If the world moves to clean energy investors would suffer major losses. The loss in demand for carbon-based energy (oil, gas, coal), naturally, makes stakeholders in these companies defensive and strategic about their survival. The economic stake investors have is a reason that influences objections to climate change agendas.

Trump (America), Putin (Russia) and Bolsonaro (Brazil) are advocates for "big oil" and politically backed by fossil fuel companies (McKibben, 2019). Bolsonaro, of Trump, tweeted that he was offered $20 billion by pro-climate change Americans to stop the 'destruction' of the Amazon Rainforest, adding that, *"if we did not accept this offer, they would then impose serious economic sanctions on our country"* (Charner & Kottasová, 2020). It is rumored that Bolsonaro turned down the $20 billion; his government signed two executive orders to curb deforestation. One order prohibited clearing the forest by fire, a common tactic of illegal ranchers, loggers and farmers and another order authorizing an army group to patrol the Amazon to prevent banned clearing and burning operations, actions to protect the Amazon and the environment (McKibben, 2019). The rainforest plays a key role in climate change mitigation, absorbing billions of tons of heat-trapping carbon dioxide from the atmosphere each year. Its vast acreage of trees serves as an umbrella and an "air conditioner" for the planet, influencing global temperature and rainfall patterns (Charner & Kottasová, 2020)..

Money Market Funds either see the writing on the wall or have a sense of moral consciousness because they divested more than eleven trillion dollars of their fossil-fuel holdings. Peabody Energy, the largest American coal company, filed for bankruptcy in 2016. It cited divestment as one of the reasons weighing on its business downturn. Shell Oil Company stated that divestment had a "material adverse effect" on its performance (McKibben, 2019). Religious institutions, including the Vatican, have divested from oil and gas. Pope Francis summoned industry executives to the Vatican requested they abandon carbon-based energy. Yet, despite the rising mixed financial sentiment regarding fossil fuel investment, In the three years since the U. S. ended participation in the Paris Agreement, JPMorgan Chase, America's largest bank has reportedly committed a hundred and ninety-six billion dollars in financing for the fossil-fuel industry to drill for new oil finds (McKibben, 2019). Much of corporate America failed to adequately support policies that addressed the threat of climate change. Many individual corporations, professed to be on the side of fighting climate change, for the sake of public image. Hypocritically, however, they supported the American Petroleum Institute, who are adversaries of climate-change actions and proponents (Senators Sheldon (D-RI) and Chuck Schumer, Chuck (D-NY).

"Climate change impacts are already costing the federal government money, and these costs will likely increase over time as the climate continues to change" according to a nonpartisan, Government Accountability Office (GAO) report. The GAO, which is seen as the watchdog for Congress, has spent two years researching and writing the report, which will make brutal reading for a climate denying President (Rowell, 2017). The GAO indicates that the Federal US Government has spent over $350 billion over the last decade paying out for extreme weather and fire events. These costs will likely increase as climate change increases and could reach over $100 billion per year by the end of the century. This could be an underestimate, with losses in labor productivity as much as $150 billion by 2099 alone (Rowell, 2017). The report details 2017 costs associated with climate change in the US, including $205 billion for domestic disaster response and relief; $90 billion for crop and flood insurance; $34 billion for wildland fire management; and $28 billion for maintenance and repairs to federal facilities and federally managed lands, infrastructure, and waterways. The GAO criticizes the response to climate change by the "climate denying" Trump Administration, which continues to insult and assault the science of climate change. Scientists within federal government working for the Trump Administration on climate change have dismantled every climate change initiative undertaken by the Obama Administration (Rowell, 2017). The report was written at the request of Republican Senator Susan Collins and Democratic Senator Maria Cantwell, the lead Democratic on the Senate energy committee, who represents Washington, one of the states affected by wildfires.

> According to Cantwell *"My colleagues no longer have to take it from me, the Government Accountability Office tells us climate change will cost taxpayers more than a half a trillion dollars this decade, and trillions more in the future unless we mitigate the impacts" (Rowell, 2017).*

Senator Cantwell stated, *"We need to understand that as stewards of the taxpayer that climate is a fiscal issue, and the fact that it's having this big a fiscal impact on our federal budget needs to be dealt with."* Susan Collins, the Senator for Maine, added: *"Our government cannot afford to spend more than \$300 billion each year in response to severe weather events that are connected to warming waters, which produce stronger hurricanes." "We cannot ignore the impact of climate change on our public health, our environment, and our economy,"* says. *"This nonpartisan GAO report Senator Cantwell and I requested contains astonishing numbers about the consequences of climate change for our economy and for the federal budget in particular."* If politicians do not believe in the science of climate change, nor the economics, what will it take to convince them act? Conventional knowledge believes politicians will procrastinate actions to mitigate climate change or never act with bipartisan decisive effectiveness. Political interest is focused on money they can put in his pocket and in the pocket of their wealthy supporters. The only way to address climate change with any hope of improvement is to vote for political candidates who support climate change and politicians who do not, vote them out of office. Many politicians think climate change is a "hoax, fake news."

Few countries are willing to follow the example of Greenland. Greenland's west coast is estimated to contain about 18 billion barrels of oil, according to a recent study from the Geological Survey of Denmark and Greenland. The U.S. Geological Survey has previously estimated that there may be double that volume in crude and natural gas in the east (MORTEN BUTTLER/BLOOMBERG, 2021; API, 2021)). However, Greenland banned all future oil explorations, citing concerns for the climate. This is an action taken by Greenland to protect their environment but their demand and need for fossil fuels did not change. Eighty percent of Greenland's oil imports come from Sweden, though it also sources oil from Iceland, Netherlands, Denmark, and Belgium to meet its petroleum needs; in 2020 Greenland imported nearly $80 million in refined petroleum. While Greenland is willing to forego their economic advantages by banning oil exploration because other industrial countries are not taking such drastic measures the world environment is still impacted by climate change / global warming. The icecaps in Greenland are melting causing the worldwide water levels to rise. While Greenland sacrifices its oil economic potential because of concerns for the climate, they still are affected because other countries have a different energy policy. The problem of global warming / climate change will not be solved until all countries in the world convert to clean energy and eliminate fossil fuels as an energy source.

Chapter X

Water Pollution in America

One can easily argue that water is the most critical natural resource in the world. Water is critical to sustaining life. The average American consumes 1 to 2 liters of drinking water per day. Practically, all drinking water in the United States comes from fresh surface waters and ground water aquifers (EPA, 2022). The earth is 70% water (Water Science School, 2019A). The oceans contain about 96.5 percent of all Earth's water. Approximately 60% of the human body is water. According to H.H. Mitchell, *Journal of Biological Chemistry 158,* the brain and heart are composed of 73% water, and the lungs are about 83% water. The skin contains 64% water, muscles and kidneys are 79%, and even the bones are watery: 31% (Water Science School, 2019B). Each day humans must consume a certain amount of water to survive. Of course, this varies according to age and gender, and also by where someone lives. Generally, an adult male needs about 3 liters (3.2 quarts) per day while an adult female needs about 2.2 liters (2.3 quarts) per day. All of the water a person needs does not have to come from drinking liquids, as some of this water is contained in the food we eat.

Water serves a number of essential functions that keep us all going

- A vital nutrient to the life of every cell, acts first as a building material.
- It regulates our internal body temperature by sweating and respiration
- The carbohydrates and proteins that our bodies use as food are metabolized and transported by water in the bloodstream;
- It assists in flushing waste mainly through urination
- Acts as a shock absorber for brain, spinal cord, and fetus
- Forms saliva
- Lubricates joints
- Used for washing and bathing (our bodies, our clothes, our homes, our cars and more)

The Clean Water Act was made law 50 years ago. Since that time the objectives identified in the law have fallen far short of the goals and present a grave danger to our environment. Researchers argued that the Environmental Protection Agency (EPA) has outdated standards and requirements for clean water conditions. The needs to update water assessment regulations and according to a report by the Environmental Integrity Project (EIP) approximately one-half of the rivers, streams, and lakes in the United States are too polluted for swimming, fishing, or drinking.

- "A new report by the EIP has found that 50 years since the passage of the Clean Water Act, the country's waterways are severely polluted."

- "The report found about half of the river and stream miles and lake acres across the U.S. are too polluted for swimming, fishing or drinking."
- "Researchers argued the Environmental Protection Agency needs to update water assessment regulations and allocate more funding for staff and resources."

The EIP has found that the EPA has been neglectful in its responsibility to protect America's natural water resource. EIP found that half, 51% of the700, 000 miles of America's waterways that were studied, including rivers and streams miles, are polluted. Across the U.S., EIP projects 27 percent of America's rivers and stream miles have been assessed during the most recent cycle. The same goes for 51 percent of lake acres and 76 percent of bay areas. The EPA delegates responsibility for controlling a clean water environment to the States.

The main causes of Water pollution are: (ECU, 2019; Nova Biologicals Team, 2018)

- Rapid Urban Development
- Improper Sewage Disposal / Marine Dumping
- Fertilizer Run-Off.
- Oil Spills.
- Chemical Waste Dumping. / Industrial Waste / Agriculture chemicals
- Radioactive Waste Discharge.
- Global Warming
- Another listing of the causes of Water Pollution can be found in Appendix IV. (Environmental Pollution Centers, 2022)

Reduce the Risks of Water Pollution

Corporations are responsible for much of the pollution of our waterways, however, there are actions individuals can take to reduce the risk of water pollution and keep the water systems clean. (ECU, 2019; Nova Biologicals Team, 2018)

- Never pour fat drippings (or any type of grease) down your drain or disposal.
- Never pour household chemicals down the drain or toilet/ do not flush trash.
- Do not use your toilet as a wastebasket. Avoid flushing anything down the toilet except toilet paper.
- Never flush old medications down the toilet.
- Use minimal detergent, or opt for an *earth-friendly* brand that is free from phosphates and sulfates.
- Minimize the use of pesticides, fertilizers, and herbicides. Never dispose of these down a nearby sewer drain.
- Pick up after your pets
- Maintain your automobile in good condition.

Water pollution laws are essential in the entire scheme of protecting the environment. Human behavior that pollutes the air and water is tantamount to committing suicide. The actions of corporations and companies who pollute the air and water but are astute to protect their personal and family air and water environment from pollution are committing acts tantamount to homicide.

CHAPTER XI

What can be Done – Solutions

A new approach and different strategies are needed to address climate change in ways that motivate people and politicians to develop a sense of urgency in addressing the problem. No doubt, human thinking and human behavior must change. It is empirically accepted that human behavior is the primary cause of the earth's warming response. Global warming is the earth's response to human stimulus. The psychology of people internationally needs to be reconstructed to address the climate issues in terms of the long-term perspective. The logistics of implementing Climate Change and Global warming solutions are subsequent to constructing the minds and will of humans to behavior environmentally positive. Showing respect for the environment should begin with the education process starting with pre-school, elementary and continues through secondary and college education. Energy conservation and environmental protection should be second nature, like taking care of your home.

Behavior change will be challenging but critically necessary. On an individual level people need to recycle, adapt to clean energy sources, and develop behaviors that protect the environment (do not litter, make environmentally friendly dietary choices). Individuals socialized into an environmental friendly lifestyle will not have the difficulty adjusting than individuals who will need to change their behavior. That is a reason economic incentives need to be given to individuals who have to undergo behavior change.

Another initiative necessary is to restructure our manufacturing processes. The environmental pollution resulting from the manufacture of products has to be eliminated. The use of fossil fuels needs to be eliminated. In other words, companies need to develop new ways to manufacturer products. If people can travel into space, then giving focus to improving the quality of life on earth should be within the realm of human beings to accomplish. The transition of behavior changes should be a process that benefits individuals both, in the ease of transition, and economically. American greed is ingrained so deeply into society that people will find a way to profit from events necessary to preserve life on earth, as we know it. The profit motive should not enter into the process of saving our planet.

The development of solar, wind, water and electric energy sources need to be accelerated and established as the mainstream source of energy, worldwide.

Think of climate change in relationship to the coronavirus (Covid-19) pandemic. America made many concessions to mitigate the Covid-19 pandemic (free testing, free shots, PPP allotments, Rental subsidies, economic debt deferments). There are also similarities between human response to climate change and their response to Covid-19. Some people refuse to comply with Covid-19 guidelines. Many of the people, who reject the guidelines (mask, distancing, washing hands, and confinement) for health purpose, got sick. Many died. Similarly, people die due to polluted air and water. Both can have a devastating impact on human life. Both are conditions caused by humans. The point is that humans tend to neglect responsibility in the midst of tragedy when they must change or modify their routine behavior. That human disregard for obeying lifesaving guidelines is ignored, only speaks to the disregard people have for other people and subconsciously for themselves. It speaks to the human value system. The lack of human perspective. It magnifies the evolution of a selfish contemporary culture with a negligent disregard for the planet, a result of which has been to create the condition of global warming. Humans have a self-destructive tendency. Extending the human destructive nature outside of the human being is easy. If a person has diminished regard for themselves and other people, disregard for nature is an extension of a diminished regard for others. If America handled the coronavirus responsibly, fewer people would have been infected and fewer would have died. Besides family, close friends and contributory associates, humans don't seem to have a well-defined disposition for caring about the broader segments of the human race, other people outside their immediate sphere and some people do not even care about those close to them. Racism is one example of that concept. Income inequality is another example. The saying, *"every man for himself and God for us all"* is a relevant description of the 21st century human consciousness. It is a proven fact that failure to follow the advice of science regarding Covid-19, resulted and results in increased infections and deaths.

The Covid-19 pandemic gives us an insight into our behavior toward climate change. The Covid-19 crisis is an immediate threat. As of September 2020, there are 6,874,982 infections and 200,275 American deaths from coronavirus. Worldwide, there are 31,800,000 infections and 974,000 deaths. As of 2021, there have been over 800,000 Covid-19 deaths in America. It is unfortunate the United States pulled out of the Paris Agreement because solving climate change requires a coordinated collaborative international effort by all industrial countries. The US is a significant polluter and was a world leader in the fight against climate change before Trump got in office. Thank goodness, he was defeated by Joe Biden in 2020. The concern for clean air and climate change alone, should be enough to never want "Trumpian" politicians in office. The coronavirus would be better handled if the effort was more effectively coordinated worldwide. The US has approximately 20%-25% of the coronavirus deaths and 3%-4% of the world population. Considering how the US is handling the coronavirus pandemic gives us insight into why the climate change problem is stagnant, at least in the US. The Trump presidential administration sent mixed communication and false information to the American public about Covid-19. The Biden Administration is supportive of climate change but let's face it. The only way to save the earth is to completely eliminate fossil fuels that pollute the earth and to adopt clean energy behavior. Based on the American socialized mentality, this is a big challenge.

The Trump Administration	**Science Experts**
Don't enforce wearing mask.	Strongly recommends wearing mask.
Don't enforce crowd control.	Strongly recommends at least 6ft. distancing.
Encourages school openings.	Strongly recommends safe School strategy.
Encourages back to work.	Strongly recommends cautious return to work.

The problem is that despite the threat and life-threatening danger of Covid-19, a percentage of Americans still put themselves and others at risk. The presidential leadership of our country and science experts are not in total agreement. We have more expert professionals in America that were in polar disagreement with the Trump Administration position on Covid-19 strategy than anywhere else in the world. The same situation exists with climate change (Appendix I). The Trump Administration was in polar disagreement with scientific experts regarding climate change and Covid-19. In both cases mixed messages are being sent to the public. This is confusing to the public. It is difficult to develop a synchronized clear-cut coordinated plan for progress when there is no common sync in opinions, policies, decisions, behavior or agreement. As a result, inadequate response both for the Covid-19 situation and for climate change causes nothing, effective, to be done in either situation. If no real effective effort can be amassed for the Covid-19 pandemic, which is immediately threatening, and people are currently dying, it is more so understandable that efforts to address climate change are easily put on the backburner. If the concern for an immediate threat is minimized, then a threat with a distant potential to do harm is easy to ignore. This is an issue for voters to resolve. The problem of climate change, like Covid-19, must have government support. Voters need to select people who are pro-climate change and have more concern for the public welfare, period. While addressing the problem requires government leadership, it does not need to be politicized. Politics only creates confusion and disagreement, which impedes progress. This same dissonance exist in regard to Climate Change / Global Warming. Bi-Partisan consistent support is critical. Ongoing participation in the Paris Accord is also critical. The Biden administration supports climate change policies and mitigation yet at the same time the temperature change in the world is still rising. The support of the Biden Administration has not resulted in positive action to mitigate the world temperature increase.

According to NASA, responding to climate change involves two possible approaches: reducing and stabilizing the levels of heat-trapping greenhouse gases in the atmosphere ("mitigation") and/or adapting to the climate change already in the pipeline ("adaptation") NASA (2020B). One approach, advanced, to motivating people to act on climate change is to emphasize relatively simple things that individuals can do in their everyday lives. Using to this approach, instead of telling people to stop driving gas-powered cars altogether, suggest they try to take the train to work one day a week. Instead of forgoing all flights, we are asked to skip one international flight a year. The list of these everyday recommendations includes such things as bringing reusable bags to the supermarket to cut down on the amount of plastic that winds up in the earth's oceans and switching all the lightbulbs in your home from incandescent to LCD. However, these individual acts, important habits as they are to develop, are insufficient to stop climate change. Some environmental advocates believe that emphasizing simple things is contributing to climate denial. Instead of telling people to be more careful recycling soda cans we should work harder to make them understand that only the complete cessation of greenhouse gas emissions will save us (Gorman and Gorman, 2019). A few misguided politicians, mostly Republicans, believe climate change predictions are exaggerated or even fabricated, despite what science experts find (Appendix 1). More astonishing, is the lack of alarm complacency in the general public. Climate change is becoming a problem on the scale of *"climbing Mt. Everest."* It can be done but it will be a challenge. The general public needs to elect politicians who take climate change as a serious priority and ignore those who minimize the seriousness of climate change. Realistically, only coordinated large-scale political effort and human behavior change has any hope of saving civilization from the oncoming ravages of continued greenhouse gas emissions. The effort needs to be nonpartisan and non-political.

Internationally, nations must come together as they did in forging the Paris Climate Agreement (Appendix II) and agree to enforce what must be an extensive collaborative effort. Behavior change is considered inconvenient when interfacing with sources of energy and food demands. One of the most challenging solutions to implement for the success of climate change is public human behavior and commitment. It is true that replacing oil and gas with sustainable energy and switching to plant-based diets will be difficult and even painful for some. Established habits are hard to break. However, the alternative, continuing to ignore that climate change is already affecting us, will ultimately be catastrophic. It could be a matter of human survival (Gorman and Gorman, 2019). The shorted-sighted neglect, ignorance and greed of contemporary wealthy, rich partisans and the compliant politicians being bribed with campaign contributions and illegal favors, must be eliminated, before it's too late. In addition to replacing politicians in denial of climate change with politicians supportive of climate change, the voting public must realize it is not a one term or two term commitment to pro-climate change candidates. It is an ongoing commitment by voters to only support candidates who have their health, survival and best interest as a priority in their political campaigns, unlike the Trump / Pence administration. Trump, Pence and all politician who support his anti-climate change views must be eliminated from the political sphere, on an ongoing and continuous basis.

In addition, individuals must be educated in schools: elementary, secondary and at the college level to understand, learn, embrace, and actively practice behavior that supports clean air, clean water, and clean energy. This is a process of learned behavior that should take place and become institutionalized in the society over time. The current generation of individuals will be extremely challenged to embrace the behaviors supporting climate change but as the future generations are educated, behaviors supporting climate change will become normal. A positive stimulus on the part of humans will result in a positive response on the part of our planet earth.

Speaking about catastrophe, about cost and sacrifice all the time will only deter people from facing that reality. Speaking about the health benefits of low-emission lifestyles; how it is safer and better for us in terms of reducing risks to do something today will probably only have minimum consequence. Even discussion on the "coolness" of using electric bikes, driving electric cars or a smarter house that regulates temperatures and lighting controls depending on whether we are home or not, will only make an insignificant dent in the problem. The solution has to be socialized and institutionalized into the lifestyle and behavior of the society in every country. Emphasizing the health benefits and the risk improvements should be cause for people to take notice but without a process that internalizes the stability of behavior that causes people to live with energy efficiency as a way of life, any casual remedy will be short-lived.

There's enough blame for climate change to go around. Fossil fuel companies, wealthy countries, politicians, rich people and every person on earth, all own a share of the blame for climate change (Timperley, 2020). However, blaming tends to reinforce the barriers, the defenses we have. When you start to blame the older generation, their response will become defensive and negative. We should keep in mind that contaminating the air, water and soil of the earth did have benefits that introduced the world to an unprecedented standard of living. No doubt it will be a significant challenge to scroll back established ways of life, but humanity has proven itself adaptable to change. The cost of not adapting to new ways of accomplishing quality lived experiences will only result in humanity losing it all. The objective should be to move climate change efforts forward, not continue the stagnation that will keep progress from being achieved.

Psychologists and economists have long explored the conflict between short-term individual and long-term collective interests when dealing with shared resources. In other words, a significant reason for resistance to climate change is money. Existing energy companies that supply oil, gas and coal will suffer revenue loss and eventual extinction with a move to clean energy. Employees that work for the carbon gas producing industries will suffer job loss. This money-mentality short-term focus versus the longer-term focus on human survival and a healthy planet, speaks to the value system that has been socialized into the lived experience of human beings (**Anderson, 2016). People have become tolerant of experiencing the effects of natural disasters because they are able to recover, to rebuild and resume life as normal** (Deutsche Welle, 2020). That humans tend to concentrate on the immediacy of a situation rather than deal with the long-term consequences is a barrier and a dilemma to the climate change issue. The major concern is that when the tragedy makes recovery too overwhelming, will it be too late to reverse the trend of climate change, and will humanity lose the way of life it has come to know? It boils down to a change in behavior now to get ahead of the "curve" or a change in behavior later to deal with the drastic impacts of being behind the "curve." A proactive or a reactive behavior response are the human alternatives. It's the difference in taking action to control the future of humanity or waiting until it's too late and reacting to life on an earth that makes it challenging and difficult to survive.

There is precedent for how movements get started, gain momentum then change the established norm of society. Analyzing historic events that have led to social tipping points or situations that tipped the scale and encouraged the masses to follow in favor of the movement, inducing change. The bus incident in Birmingham Alabama involving Ms. Rosa Parks, a black woman, was one such situation that tipped the scale and gave momentum to the civil rights movement. When Ms. Parks refused to sit in the back of the bus in racist Birmingham, Alabama, the black community boycotted the city bus line and won reconciliations. The Vietnam War protest is another such movement that tipped the scale in favor of ending the war. Initially protestors were seen as unpatriotic, but when a photo of a girl running from the napalm bombing was shown in newspapers, almost overnight, the social tipping point happened. However, regarding global warming / climate change, if the growing number of natural disasters don't tip the scale in favor of addressing the issue with effective aggression, then we are doomed to confronting a dire situation. The truth is that the only real solution to the problem is to eliminate world dependence on fossil fuels, completely. The hardest thing for people to do is sacrifice, especially after they have experienced the advantages of what fossil fuels provide. Now we are at a point where the disadvantages are beginning to rival the disadvantages. Maybe people need to sacrifice and go without the goods and services resulting from fossil fuel production until the factories can convert to clean energy production. This is a decision that has to be made. The question is when will it be made, and it will, will it be too late?

Growing concern of the Youth

An increasing number of young students are becoming socially involved and protesting for their futures (BBC, 2019). Teenage activist Greta Thunberg scolded European MPs for failing to understand this. *"You can't just sit around waiting for hope to come,"* she said last February in Brussels. *"Then you are acting like spoiled irresponsible children. You don't seem to understand that hope is something that you have to earn."* Ms. Thunberg's points out that it takes action, not hope, to solve the climate change problem (Ortiz, 2020). It's not likely any communication experts or campaign strategists would use a 15-year-old Swedish girl with Asperger Syndrome as the frontrunner to spearhead the campaign for climate change. Thunberg is an example of the growing and concerning effort of youth who are concerned about their future and realize the current generation of decision-makers are squandering and jeopardizing the future of tomorrow's youth and adults. It is known that there is a committed minority varying from country to country between say 15 percent to 40 percent of people who are very concerned about climate change (Deutsche Welle, 2020). The youth of the world are the people with the greatest stake in climate change. Educating and mobilizing the youth in behaviors that comply with the goals of reducing global warming and energy conservation (recycling, diet consumption, consciousness when making purchases, etc.) can provide the tipping necessary to strongly motivate people to deal with climate change seriously. Youth worldwide and in America need to get involved. After all, it's their future that is at stake.

Global warming will only get worse with each year, each decade, because if for no other reason, the significant increasing population (Worldometer, 2020) will cause greater use of dirty energy (more use of fossil fuels), more pollution. The increase in world population through the years correlates with the increase in use of fossil fuels. If significant immediate action is not taken, the world may be in serious jeopardy.

Year	World Population	Yearly Change	Net Change
Change	**Net**	1.88 %	47,603,112
Change	**3,034,949,748**	1.86 %	55,373,563
1970	**3,700,437,046**	2.06 %	74,756,419
1980	**4,458,003,514**	1.77 %	77,497,414
1990	**5,327,231,061**	1.71 %	89,789,503
2000	**6,143,493,823**	1.31 %	79,254,768
2010	**6,956,823,603**	1.22 %	84,056,510
2020	**7,794,798,739**	1.05 %	81,330,639

Population growth will increase the demand for fossil fuel emissions. Unless human behavior changes, more carbon will be emerged into the atmosphere. When combined with oxygen the carbon becomes carbon dioxide (CO_2). The water on the planet absorbs gases from the atmosphere. As the earth's waters absorb CO_2, they become more acidic (Melillo et al., 2014). Fresh water life, ocean water life and drinking water can become contaminated and the challenge for human survival will be considerable.

Chapter XII

The Conclusion / Summary

What goes around, comes around. You get what you give. You get what you pay for. Crap in, crap out. These colloquial sayings are all ways of understanding the theory of Stimulus – Response when it comes to understanding that Global warming is caused by humans emitting heat-trapping gases into the atmosphere. A critical issue facing international society and the masses is realizing the urgent and critical need to focus on mitigating climate change. There needs to be alignment between the following factors: 1) to bridge the gaps between human concern / agreement, 2) motivation to act, 3) acceptance that behavior needs to change and 4) the realization that urgency to act is critical. Nature and the environment are only responding to the behavior of human beings in the world.

Climate Change is one of the defining issues of this era. This generation's legacy will be determined based on how the issue of climate change is addressed. Changing weather patterns that threaten food production, wildfires that destroy the ecology, rising sea levels that increase the risk of catastrophic flooding, devastating storms that disrupt normalcy, are characteristic of the impact climate change has on a global scale. Without immediate action, adapting to the disastrous impacts in the future will be more difficult and costly. Getting people to realize and embrace the fact their actions and behaviors (stimulus) are causing the earth to respond with disastrous repercussion to humanity that will become worse without serious action.

The irony of the increasingly deterioration of earth's environmental stability is that the rich and wealthy, the politicians and those who minimize the threat of climate change having potential to destroy mankind, lack the intelligence, knowledge, leadership skills, foresight, and insight that it takes to govern and lead. Their leadership abilities are based on selfish motives. The reality is that resolution of the climate change problem rest with the people, the voters, in democratic industrial countries. Voters need to support candidates who have sincere unselfish concern for the future of their respective civilizations. When the correlation is made between

1) The evolutions of industrialization, 2) The increase in population, 3) the amount of fossil fuel emissions emitted as industrialization increased, 3) the increase of natural disasters over the years and 4) the rising temperature in the earth's atmosphere; there is direct correlation between the increases of all variables mentioned.

Earthquakes, hurricanes, tornadoes, floods, wildfires, drought are considered by some to be acts of God. In reality, these disasters are responses by earth to the stimulus provided by humans. Humans cause the problem, and it is within the power of humans to correct the problem before it's too late. Framing climate change in a religious context or attributing any blame on God is absurd. God is not destructive. God is constructive. God has given humans the knowledge to correct the problem created by humans, themselves. Just as humans make excuses to deny the problem, they make excuses for the cause of the problem and they make excuses for occurrences of the problem. This is a human dilemma that needs to eliminate the ignorant confusion. That humans ignore the earth's devastating response to the abuse of earth's resources cannot be attributed to God. Placing the blame on God, takes it out of the human realm to correct. Humans are just stupid, in terms of being blindly led by greed which causes them to willfully ignore the signals and signs of earth's response to their actions. Shifting blame is a form of denial. Another religious argument is that God gives humans "free will," even if it's the free will to destroy themselves. Ironic. God cannot be held accountable for the actions of humans. Any religious argument that rationalizes the earth's response to climate change is ridiculous. The earth experiences and enacts disastrous events (response) because of the actions (stimulus) and behavior of humans. Attributing any responsibility to "acts of God" only takes the responsibility for addressing and resolving the problem outside of the human realm. It's irresponsible, cowardly and a lazy smokescreen for doing nothing.

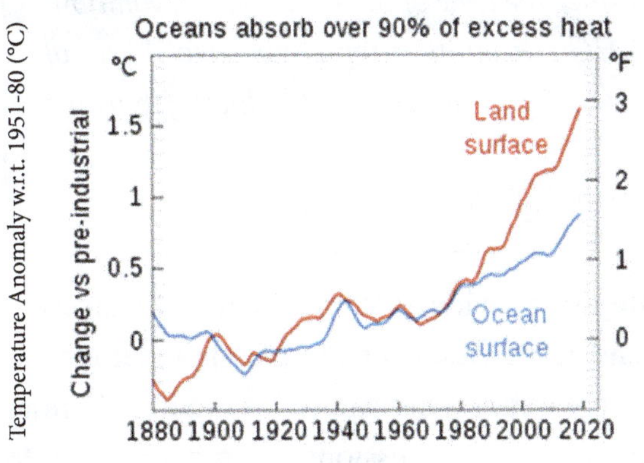

The reality is that over the years the earth's temperature has become warmer, both the land and water temperatures. The temperature anomalies (vs. 1951-1980) averaged over the Earth's land area and sea surface temperature anomalies (vs. 1951-1980) averaged over the part of the ocean that is always free of ice (open ocean). NASA data shows that land surface temperatures have increased faster than ocean temperatures (NASA, 2020A).

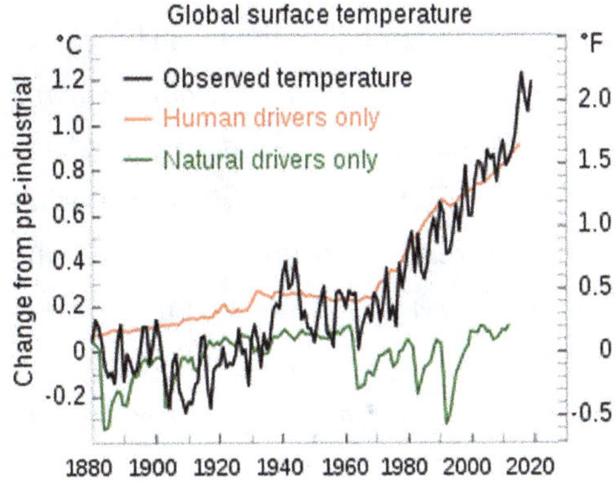

It is also a proven fact that the major cause of global warming is human behavior.

Observed temperature from NASA versus the 1850–1900 average as a pre-industrial baseline. The main driver for increased global temperatures in the industrial era is human activity, with natural forces adding variability (Knutson et al. (2017).

> Here are some basic well-established scientific facts (United Nations, 2019):
> *"The concentration of Green House Gas in the earth's atmosphere is directly linked to the average global temperature on Earth;*
> *The concentration has been rising steadily, and mean global temperatures along with it, since the time of the Industrial Revolution;*
> *The most abundant GHG, accounting for about two-thirds of GHGs, carbon dioxide (CO2), is largely the product of burning fossil fuels.*
> *Humans are the major cause of Climate change."*

The Fifth Assessment Report provides a comprehensive evaluation of the rise in sea levels, and its causes, over the past few decades. It also estimates cumulative CO_2 emissions since pre-industrial times and provides a CO_2 budget for future emissions to limit warming to less than 2°C. The report found that half of this maximum amount was already emitted by 2011 (United Nations, 2019):

> *"From 1880 to 2012, the average global temperature increased by 0.85°C.*
>
> *In October 2018 the IPCC issued a* special report *on the impacts of global warming of 1.5°C, finding that limiting global warming to 1.5°C would require rapid, far-reaching and unprecedented changes in all aspects of society. With clear benefits to people and natural ecosystems, the report found that limiting global warming to 1.5°C compared to 2°C could go hand in hand with ensuring a more sustainable and equitable society. While previous estimates focused on estimating the damage if average temperatures were to*

rise by 2°C, this report shows that many of the adverse impacts of climate change will come at the 1.5°C mark.

A number of climate change impacts that could be avoided by limiting global warming to 1.5°C compared to 2°C, or more. For instance, by 2100, global sea level rise would be 10 cm lower with global warming of 1.5°C compared with 2°C. The likelihood of an Arctic Ocean free of sea ice in summer would be once per century with global warming of 1.5°C, compared with at least once per decade with 2°C. Coral reefs would decline by 70-90 percent with global warming of 1.5°C, whereas virtually all (> 99 percent) would be lost with 2°C.

Oceans have warmed, the amounts of snow and ice have diminished, and the sea level has risen. From 1901 to 2010, the global average sea level rose by 19 cm as oceans expanded due to warming and ice melted. The sea ice extent in the Arctic has shrunk in every successive decade since 1979, with 1.07 × 106 km² of ice loss per decade.

Given current concentrations and ongoing emissions of greenhouse gases, it is likely that by the end of this century global mean temperature will continue to rise above the pre-industrial level. The world's oceans will warm, and ice melt will continue. Average sea level rise is predicted to be 24–30 cm by 2065 and 40–63 cm by 2100 relative to the reference period of 1986–2005. Most aspects of climate change will persist for many centuries, even if emissions are stopped."

It is generally accepted that climate change is real and happening and driven in some way by human behavior. An overwhelming majority of climate scientists (97%) agree that climate change is caused by human behavior (Appendix I). Despite the large numbers of people accepting that climate change is real and driven by human behavior, there is a 'consensus gap.' The consensus gap is the difference between the public's perception of how much agreement there is among scientists that humans are causing global warming (typically about 50%), compared to the actual 97% consensus among scientists publishing in the peer-reviewed literature (APS, 2020). Major worldwide Scientific Academies, Societies, and Associations have strongly endorsed the reality of climate change, global warming and the need for immediate corrective action Gleick, 2017).

One of the ways of dealing with this "consensus gap" is to raise awareness of the scientific consensus on climate change. The importance of this cannot be overstated. Research shows that people are more likely to support policy actions to reduce carbon dioxide emissions if they are aware of the overwhelming agreement among experts that we are causing global warming (Appendix 1). Studies of climate change perceptions in Australia, the UK and the US show that only very small numbers of people deny that climate change is happening. The figures range from between 5 to 8% of the population. However, this small minority can be influential in casting doubt on the science, spreading misinformation and impeding progress on climate policies. Science denial can be stopped by first explaining the psychological research into why and how people deny climate science. The best way to neutralize misinformation is to expose people to a weak form of the misinformation. First, explain the fallacy employed by the myth. Once people understand the techniques used to distort the science, they can reconcile the myth with the fact. However, if a consensus regarding climate change developed, the world community will need to accept the challenge of behavior change (APS, 2020; Van Boven, Ehret & Sherman, 2018).

The Landmark Paris Agreement (Appendix II)

On December 12, 2015, 195 nations, Parties to the United Nations Framework Convention on Climate Change (UNFCCC) reached a landmark agreement, adopting the Paris Agreement to **decarbon-**

ize the environment, collaborate to reduce global greenhouse gas (GHG) emissions and limit global temperature increases. The Paris Agreement goals are to achieve the following: (Mantle314, 2015)

- To accelerate and intensify the actions and investments needed for a sustainable low carbon future.
- Restrict the increase in global average temperature to well below 2°C above pre-industrial levels and pursue efforts to limit temperature increase to 1.5°C above pre-industrial levels.
- Parties agreed to a goal to peak global GHGs as soon as possible and achieve carbon neutrality or net-zero emissions by the second half of this century.
- Increase adaptive capacity and foster resilience.
- Increase financing of low carbon and climate resilient development for poor countries.

While there are still those who resist actions to mitigate climate change and who claim climate action is too difficult, or inconsistent with economic stability and growth, their voices are becoming muted by the growing support in favor of combatting climate change. The Paris Agreement transcends conventional international relationships and, for the first time, brings all nations into a common cause to undertake efforts to combat climate change and adapt to its effects, to also help align developing countries to do the same. On Earth Day, 22 April 2016, 175 world leaders signed the Paris Agreement at United Nations Headquarters in New York. This was by far the largest number of countries ever to sign an international agreement on a single day. There are now 186 countries that have ratified the Paris Agreement. A monumental potential benefit of the Paris agreement is that is demonstrated the ability of all nations in the world to come together with a collaborative mutual effort to resolve common interest. This effort can be a catalyst for world peace, solving worldwide hunger and health problems and generally foresee hope of living in a world where people get along.

Motivated by irresponsible malice and envious jealously to undo the accomplishments of his predecessor, President Barak Obama, on June 1, 2017, Donald Trump withdrew the United States from the Paris Agreement. Almost 200 countries signed the Paris Agreement in 2015 and committed to reduce greenhouse gas emissions. Each country set its own goals, and many wealthy countries, including the US agreed to help poorer countries pay for the costs associated with climate change. The US is the only country to pull out of the pact. Trump's withdrawal from the agreement in 2017, uncommitted the United States from actions designed to reduce greenhouse gas emissions, and obligation to contribute to the international climate fund for poorer nations. The US initially committed to reduce national greenhouse gas emissions by approximately one quarter by 2025, compared with 2005 levels. The US is not on track to achieve that goal. Subsequent to withdrawal, the Trump administration has ignored federal commitment limits on carbon emissions, including rules about how much pollution can be emitted by power plants, cars and trucks. The US is the second largest carbon emissions polluter in the world behind China. India ranks third. Without US participation, reaching global warming goals may be unlikely. Another assumption one can make from the actions of Donald Trump is that individuals with his similar mentality are not suitable to lead in a world that hopes for harmony. A new administration can reverse the Trump administration withdrawal. If a future administration rejoins the Paris Agreement hope can be restored that our planet and inhabitants can be saved (HERSHER, 2019).

Bob Irvin, President of American Rivers, upon Trumps withdrawal from the Paris Agreement, issued the following statement: (Kober, 2017)

> *"In withdrawing from the Paris Agreement, President Trump is abdicating the responsibility of the United States to be a global leader and is putting our nation's health, economy and national*

security at risk. Far from making America great again, his decision will leave our planet worse for future generations."

"Climate change is here now and is already impacting our rivers and clean water supplies. Without bold action to stop climate change, more severe floods and droughts, more waterborne diseases, and increasingly scarce water supplies will threaten communities in the United States and around the world."

"The president's head-in-the-sand approach to climate change contrasts sharply with the leadership and moral courage of countless individuals, businesses and cities that are working tirelessly to stop dirty fossil fuel pollution and strengthen communities against climate impacts including increased flooding and drought."

"American Rivers will continue to stand with these local leaders, and we will continue helping communities build their resilience with innovative river conservation solutions. We will work to ensure the United States remains a global leader in river restoration and protection, because a healthy river is a community's best defense against the impacts of climate change."

The American Rivers organization protects wild rivers, restores damaged rivers and conserves clean water for people and nature. Since 1973, The American Rivers organization has protected and restored more than 150,000 miles of rivers through advocacy efforts, on-the-ground projects and an annual America's Most Endangered Rivers® campaign. Headquartered in Washington, DC, American Rivers has offices across the country and more than 250,000 members, supporters and volunteers" (Kober, 2017).

As a matter of fact, climate change was left out of Trump's 138-page federal budget document, requesting Congress to approve a total of $4.8 trillion to fund the president's priorities across the federal government in fiscal 2021 (Chemnick & Frank, 2020). The agencies that are most responsible for addressing climate change at home and abroad, including EPA and the departments of Energy and State, received steep cuts under Trump's proposed budget. Programs in those agencies that deal directly with climate change will have the biggest cuts. EPA, which has existed for half a century would receive $6.7 billion for fiscal 2021, a 27% cut compared with enacted levels. It also would also have to cut its workforce by approximately 11%, to the lowest levels since 1985 (Chemnick & Frank, 2020). It's obvious that Trump's intentions are to dismantle the US clean environmental effort in favor of fossil fuels. Hopefully, Trump's irresponsibility will be evident to voters in 2020 in a way that causes them to vote Trump out of office.

A major problem confronting climate change is that it is a universal problem that is caught up in politics, partisan politics, and special interest. Politicians and individuals who can make a difference in re-establishing a natural order to the universe, but do not, are irresponsible to ignore climate change. A greater jeopardy is that their irresponsibility to neglect climate change will influence significant numbers of the general public to do the same. It seems they would rather risk universal order than comply with the principles of universal order. This fact is evident by the actions taken by humans' failure to respond in a positive manner to the earth's response to human pollutant behaviors. Chief among the violators was the defeated President of the United States, Donald J. Trump. He is a man who is more concerned about money than people. His intelligence level and IQ regarding climate change is considerably less and below that of 17-year-old Greta Thunberg, who is an astute young lady when it comes to caring for the universe and for people. Donald Trump should take lessons from her, become her student. He has much to learn.

Data shows a direct correlation between the increase industrialization, increasing population growth, and increasing fossil fuels emission into the environment (atmosphere) which in turn shows a correlating increase in natural disasters throughout the world. The increase in industrialization over the years, corresponds with a direct increase in fossil fuel emissions, which also directly corresponds to the increase in the occurrence of natural disasters throughout the world. It does not take a "rocket scientist" to connect the dots in the relationship between industrialization, population growth, fossil fuel emissions, world temperature increase, and natural disaster occurrences.

A welcome response to global warming and climate change is the emerging manufacturing of electric cars. Companies like Tesla and Lucid produces cars that run exclusively on electricity. In the early stages of the development of energy efficient cars, companies such as General Motors and Ford developed what they labeled as Hybrid Cars, cars that ran on electricity and gasoline. The driver could switch to either fuel source when necessary. However, this was determined to still not address the problem of eliminating the use of fossil fuels because the electric energy source had limits to the mileage usage. Now most major automobile manufacturers are coming out with 100% electric energy cars.

The Intergovernmental Panel on Climate Change (IPCC) issued a February 27, 2022, report on the current status of global warming and climate change. The report was approved by 195 countries. A major finding in the report is that people around the world need to take climate change / global warming very seriously. Mitigating climate change involves behavior change. Either behavior change will occur voluntarily, or it will occur without choice and will involve adaptive behavior, even behaviors necessary to survive. Included in the report were a listing of frequently asked questions (FAQ's) taken directly from the IPCC report that can be found in Appendix III (Working Group, 2022).

Summarizing the Climate Change / Global Warming Issue in Graphs

The correlation between the global population increase, the increase in industrialization and the increase in fossil fuels polluting the earth with the increase in global temperature and natural disasters is without doubt cause to take notice and induce action. Correlation between the following factors: 1) Increase in Industrialization, 2) Population increases 3) Increase in fossil fuel emissions, 4) Increase in global temperature, and 5) Increase in natural disasters will present a logical explanation of the global warming / climate change phenomena.

Increase in Industrialization

The increase in industrialization is attributed to the following factors; the Industrial Revolution itself, including: the emergence of capitalism, European imperialism, efforts to mine coal, and the effects of the Agricultural Revolution. The Industrial Revolution began in the 1830s to 1840s in Britain, and soon spread to the rest of the world, including the United States.

This transition included going from hand production methods to machines; new chemical manufacturing and iron production processes; the increasing use of water power and steam power; the development of machine tools; and the rise of the mechanized factory system. Output greatly increased, and a result was an unprecedented rise in population and in the rate of population growth. The textile industry was the first to use modern production methods, and textiles became the dominant industry in terms of employment, value of output, and capital invested.

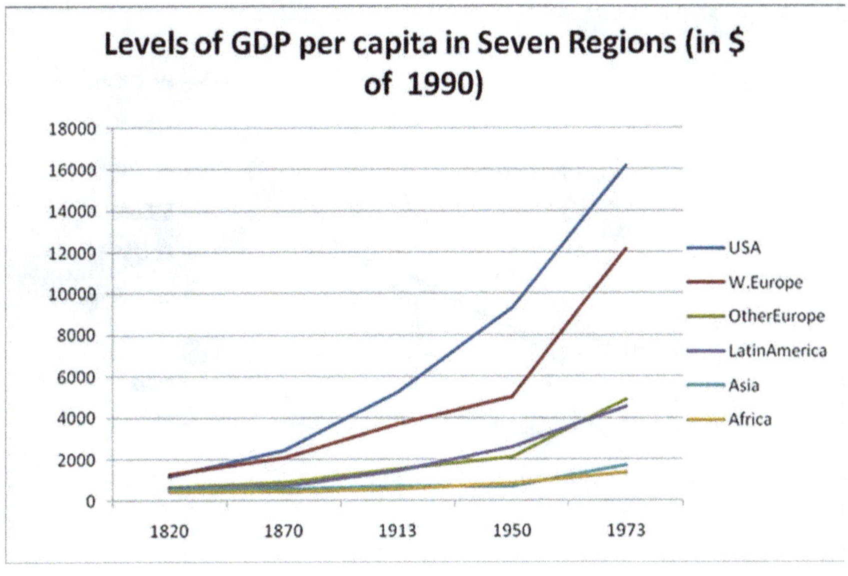

(Tilly, 2018)

The industrial revolution, caused an international transformation in the increase of real income per person in England and, throughout the world. As the Western world industrialized, historians argue the industrial revolution was one of the most important events in history, marking the rapid transition to the modern age. However, they are divided over how the industrial revolution affected ordinary people, often called the working classes. One group, the pessimists, argues that the living standards of ordinary people fell, while another group, the optimists, believes that living standards rose. All agree that the industrialist entrepreneurs gained the real and greatest economic benefit of the industrial revolution (Nardinelli).

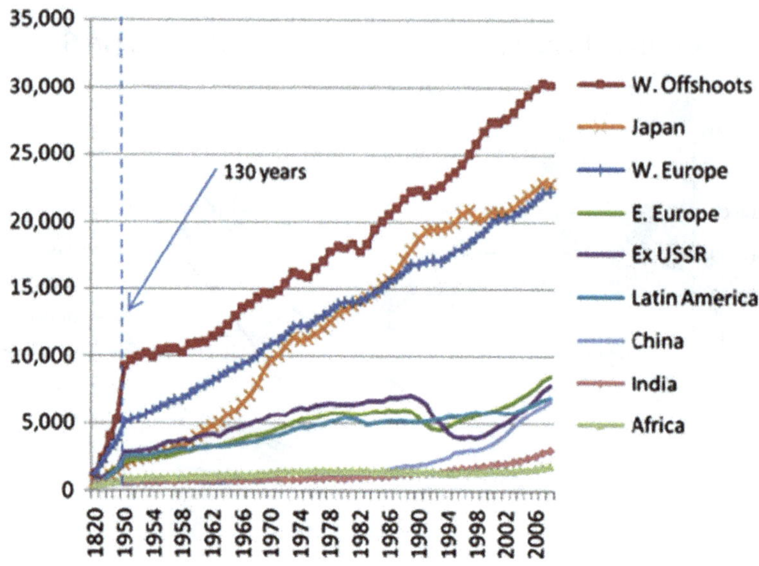

Yifu Lin, Justin and Rosenblatt, David (2013)

The Industrial Revolution shifted the production of goods from an agrarian economy to a manufacturing economy where products were no longer made solely by hand in sparsely populated areas, but product manufacturing moved to urban areas and were produced by machines. This led to increased production and efficiency, lower prices, more goods, improved wages, and migration from rural areas to urban areas. Industrialization is driven by a combination of factors including government policy, labor-saving inventions, entrepreneurial ambitions, and the demand for goods and services. One must note that the increase in population is a significant factor in the demand for goods and services. Also as changes in perspectives, education, and cultural phenomena changes from generation to generation, likewise demand for goods and services shift in the direction of the generational demand.

Population Increase

The global population has grown from 1 billion in 1800 to 7.9 billion in 2020. "Population in the world is, as of 2022, growing at a rate of

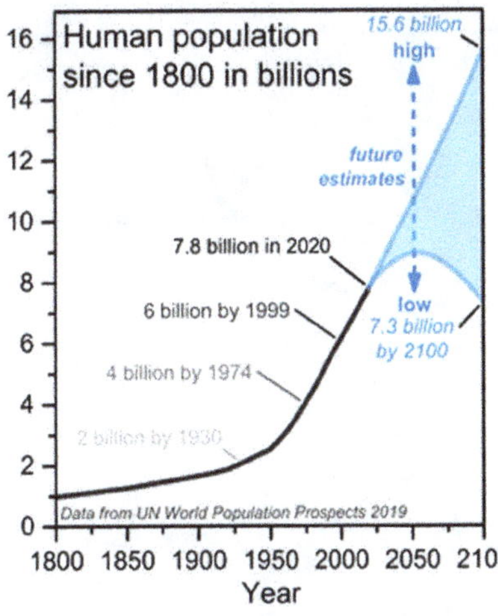

around **0.84%** per year (down from 1.05% in 2020, 1.08% in 2019, 1.10% in 2018, and 1.12% in 2017). The current population increase is estimated at **67 million people per year.** Annual growth rate reached its peak in the late 1960s, when it was at around 2%. The rate of increase has nearly halved since then, and will continue to decline in the coming years. World population will therefore continue to grow in the 21st century, but at a **much slower rate compared to the recent past.** World population has doubled (100% increase) in 40 years from 1959 (3 billion) to 1999 (6 billion). It is now estimated that it will take another nearly 40 years to increase by another 50% to become 9 billion by 2037" (WorldOMeter (2023).

Population growth increased with the industrial revolution. In **all of human history, until around 1800 it took the dynamics of the industrial revolution for world population to reach one billion.** The second billion was achieved in only 130 years (1930), the third billion in 30 years (1960), the fourth billion in 15 years (1974), and the fifth billion in only 13 years (1987). During the 20th century alone, the population in the world has grown from 1.65 billion to 6 billion. In 1970, there were roughly half as many people in the world as there are now. The latest world population projections indicate that world population will reach 10 billion persons in the year 2057 (WorldOMeter (2023).

Increase in fossil fuel emissions

The estimates come from the 2022 Global Carbon Budget report by the Global Carbon Project. It finds that the increase in fossil emissions in 2022 has been primarily driven by a strong increase in oil emissions

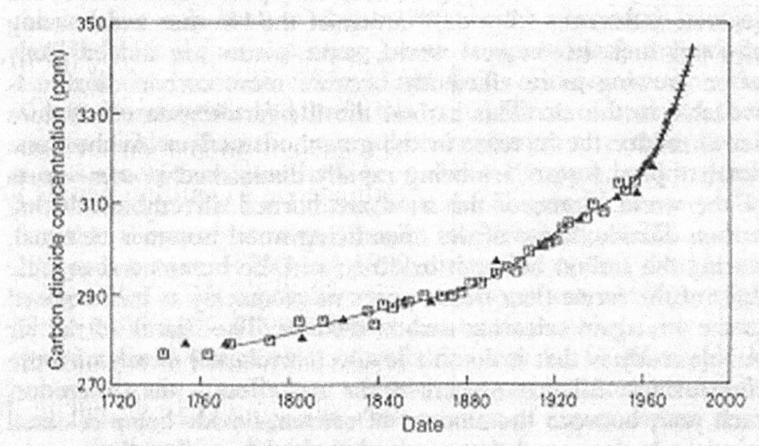

as global travel continues to recover from the Covid-19 pandemic. Coal and gas emissions grew more slowly, though both had record emissions in 2022.

Total global CO2 emissions – including land use and fossil CO2 – increased by approximately 0.8% in 2022, driven by a combination of steady land-use emissions between 2021 and 2022 and increasing fossil CO2 emissions. However, total CO2 emissions remain below their highs set in 2019 and have been relatively flat since 2015.

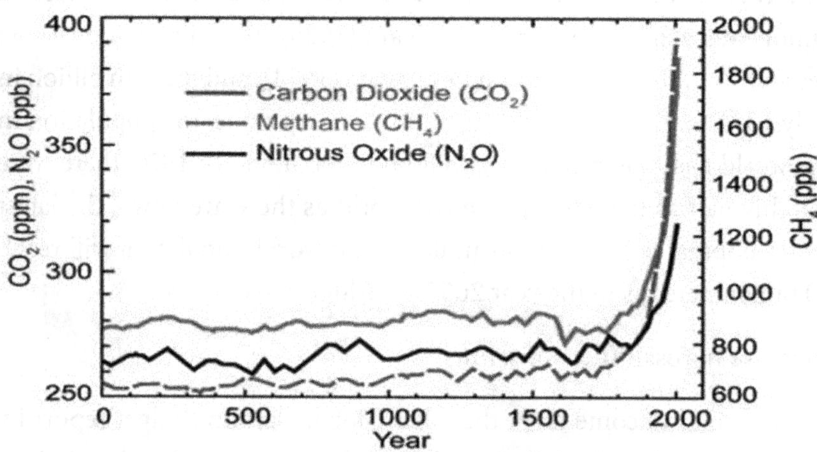

Global fossil fuel emissions primarily result from the combustion of coal, oil and gas.

Coal is responsible for more emissions than any other fossil fuel, representing approximately 40% of global fossil CO2 emissions in 2022. Oil is the second largest contributor at 32% of fossil CO2, while gas and cement production round out the pack at 21% and 4%, respectively.

These percentages reflect both the amount of each fossil fuel consumed globally, but also differences in CO2 intensities. Coal results in the most CO2 emitted per unit of heat or energy produced, followed by oil and gas.

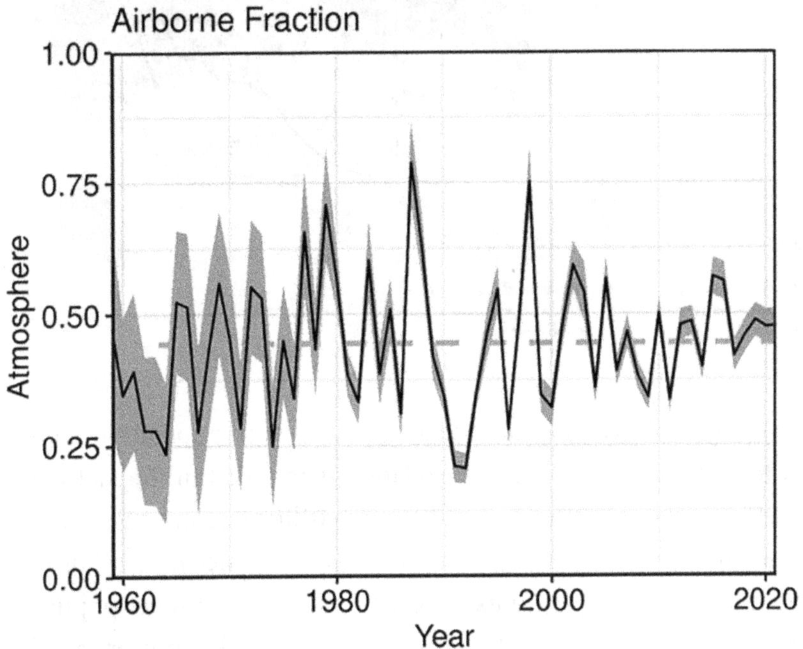

The chart above shows, the fraction of CO2 emissions that end up in the atmosphere varies from year to year. The grey dashed lines shows that around 47% of total CO2 emissions have remained in the atmosphere each year over the past decade, with the remainder being taken up by ocean and land sinks. The black line indicates the land absorption of CO2 emissions and the blue lines indicate the oceanic absorption of CO2 emissions (Hausfather, & Friedlingstein, 2022).

The relationship between Population Increase and Fossil Fuel emissions

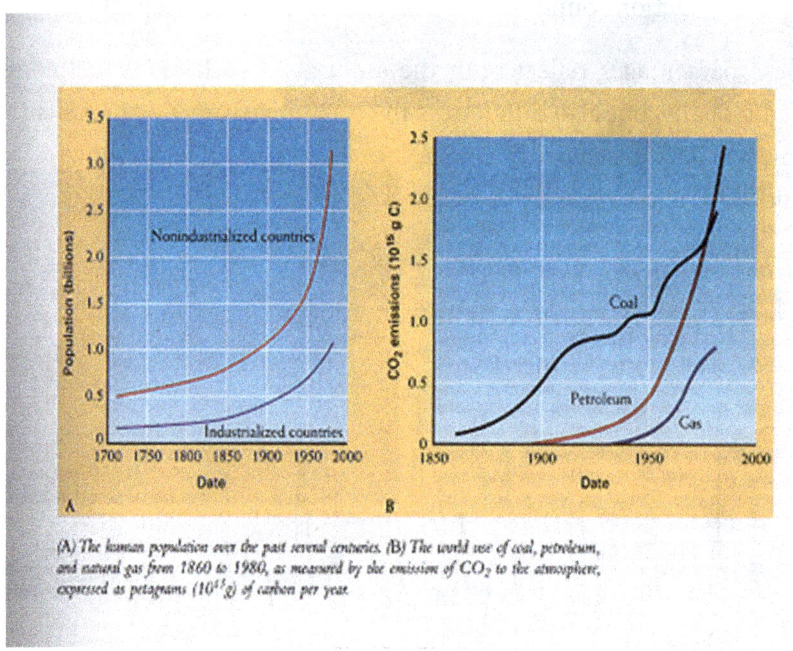

(A) The human population over the past several centuries. (B) The world use of coal, petroleum, and natural gas from 1860 to 1980, as measured by the emission of CO_2 to the atmosphere, expressed as petagrams (10^{15}g) of carbon per year.

The graphs above indicate a correlation between population increase and fossil fuel emission. The correlation is founded in the fact that as population increases the demand for goods likewise increases. The increased demand for goods causes factories to produce more, thus being reliant on increased amounts of fossil fuel energy to meet population and production demand. More fossil fuel is needed to power cars, heat homes and business, and in general to meet the increasing demand for consumer products.

Increase in global temperature

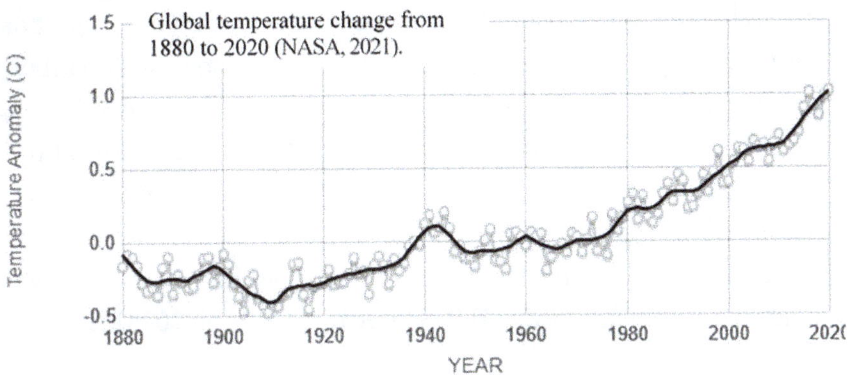

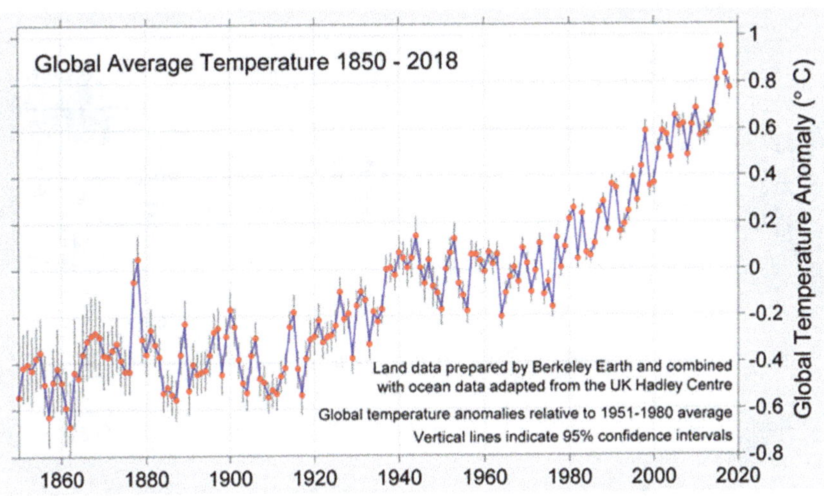

The charts above show that the world temperature has increased each decade since the 1800's, consistent with the rise of the industrial revolution. According to NOAA's 2021 Annual Climate Report the combined land and ocean temperature has increased at an average rate of 0.14 degrees Fahrenheit (0.08 degrees Celsius) per decade since 1880; however, the average rate of increase since 1981 has been more than twice as fast: 0.32 °F (0.18 °C) per decade. This period of increased temperature rise correlates with the rise in Chinese manufacturing and the United States among other countries shipping a significant portion of their manufacturing off-shore, to China and other foreign countries. The amount of future warming earth will experience depends on how much carbon dioxide and other greenhouse gases we emit in coming decades. Today, our activities, burning fossil fuels and clearing forests, add about 11 billion metric tons of carbon (equivalent to a little over 40 billion metric tons of carbon dioxide) to the atmosphere each year. Because that is more carbon than natural processes can remove, atmospheric carbon dioxide increases each year (LINDSEY & DAHLMAN, 2023).

Based on these projections, unless drastic action is taken the world will a desert melting pot with survival being the primary goal in life

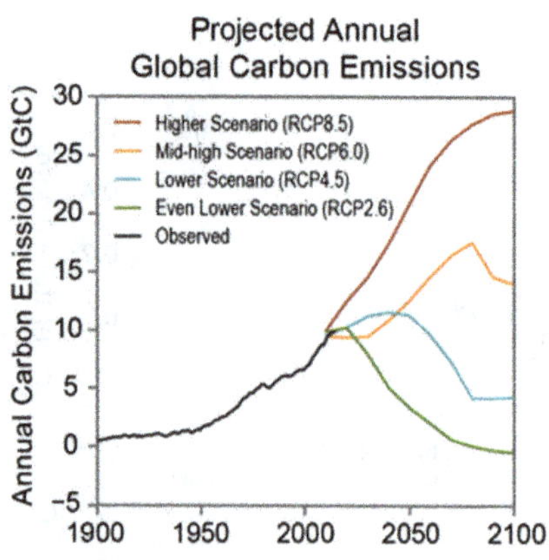

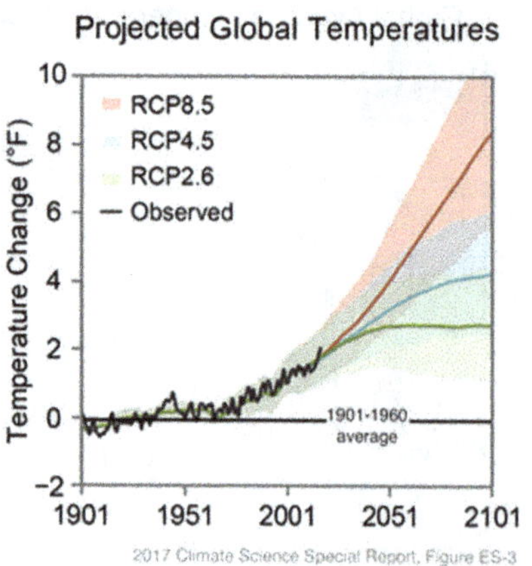

Increase in natural disasters

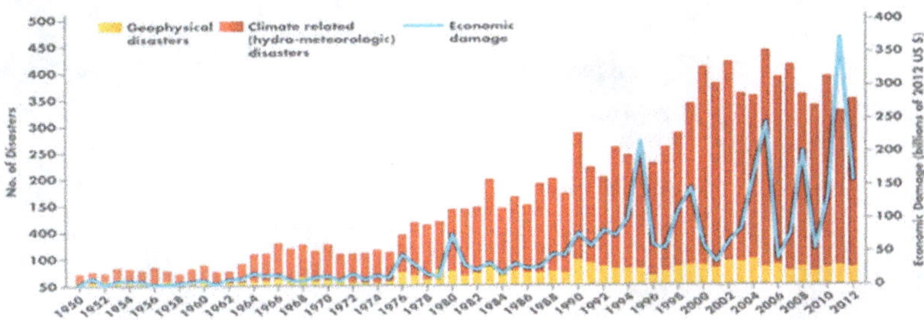

The graph above represent the increasing rise in natural disasters. The first graph shows the geophysical disasters. Geophysical disasters are disasters that are brought about by tectonic and seismic activity below the Earth's surface, such as earthquakes and volcanos. Many of these kinds of disasters have similar signs that they are about to occur, such as shaking and unstable ground. Climate related disasters refer to storms, floods, landslides and heatwaves, and have sudden and obvious effects. With slow disasters such as droughts, increases in water and soil salinity and crop losses, the impacts may take longer to emerge but they can be very severe. The graph to the right shows the trend in climate related disasters from 1900-2009.

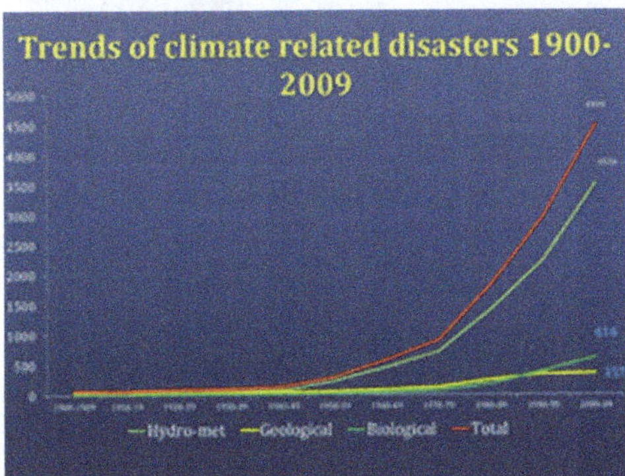

The first graph also shows damage of disasters resulting from atmospheric and geophysical in terms of economic cost. If the economic cost of the disasters caused is not enough for humans to change their behavior, and seriously address climate change and global warming crisis, then what is? When it gets to the point of humans have to be reactionary to a situation of irreversible global warming, then the standards of living we know at this point will diminish.

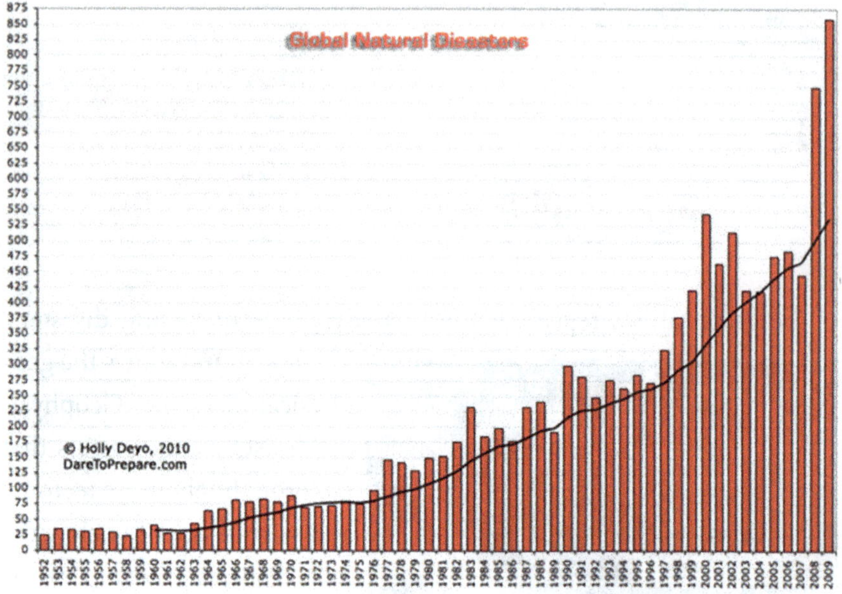

The graph above shows the trend of natural disasters that have occurred since 1952 – 2009. The increasing trend in natural disasters is obvious. What causes natural disasters to increase each year on a consistent basis? If humans cannot read the "writing on the wall" and take notice of the obvious correlation of the trends between global warming/ climate change and natural disasters, then humanity is doomed.

Natural disasters are more frequent than 30 years ago - and are costing us more

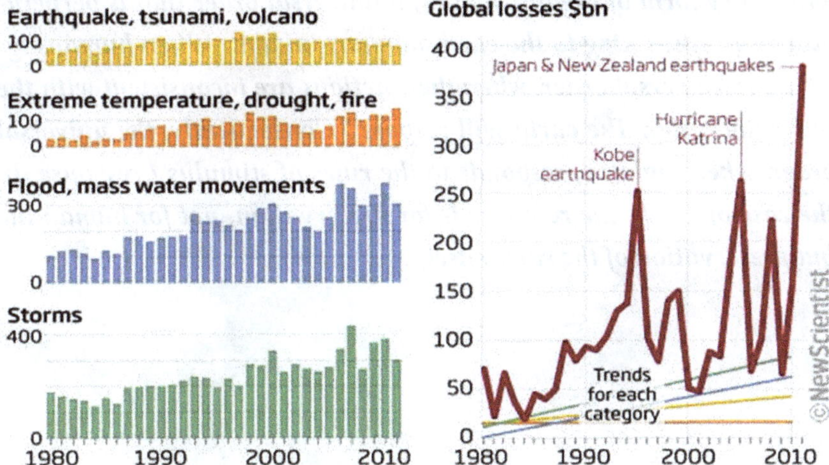

The graph above indicates the types of natural disasters that have occurred over the past 30 years and the cost associated with the destruction resulting from these disasters.

All graphs depicting the rise in Industrialization, Population, fossil fuel emissions, increase in global temperature, and the increasing trends in natural disasters skew upwards during a correlating time period. The correlation between the following factors: 1) Increase in Industrialization, 2) Population increases 3) Increase in fossil fuel emissions, 4) Increase in global temperature, and 5) Increase in natural disasters allows individuals to draw the following logical conclusions.

1. Global Warming / Climate Change is a reality
2. Global warming / Climate change is caused by fossil fuel emissions
3. Fossil fuel emissions are caused by industrialization and the population demand for goods and services.
4. The fossil fuel emission impact on global warming has caused an increase in worldwide natural disasters.

There is an order to the universe. A natural order that is consistent with every form of organic matter, a universal order that is perpetuating and sustaining to the earth humans inhabit. When humans ignore the universal order, when their actions are inconsistent with the universal order. The earth will respond. Consequently, the universal order, when violated, responds to the rules of stimulus / response. In the case of earth, the response is for preservation, not for human life but preservation of the earth, itself.

Tornadoes

Floods

Wild Fires

Hurricanes

Volcanos

Drought
Before
After

REFERENCES

Ali, Shirin (2022). About half of US water 'too polluted' for swimming, fishing or drinking, report finds. *Changing America*. Retrieved from: https://thehill.com/changing-america/sustainability/environment/600070-about-half-of-us-water-too-polluted-for- swimming/

Amruta Nori-Sarma, Shengzhi Sun, Yuantong Sun, Keith R. Spangler, Rachel Oblath, Sandro Galea, Jaimie L. Gradus, Gregory A. Wellenius (2022). Association Between Ambient Heat and Risk of Emergency Department Visits for Mental Health Among US Adults, 2010 to 2019. *JAMA Psychiatry April 2022 Volume 79, Number 4*. Retrieved from: file:///C:/Users/Public/jamapsychiatry_norisarma_2022_oi_210089_1648226517.09785.p df

Anderson, Lynne (2016). How psychology can help us solve climate change. *Published in conjunction with Oxford University's Practical Ethics blog*. **Retrieved from:** https://theconversation.com/how-psychology-can-help-us-solve-climate-change-58957

API (2021). Greenland bans all oil exploration. Retrieved from: https://www.cbc.ca/news/business/greenland-oil-1.6105230

APS (2020). The psychology of climate change denial. *Australian Psychological Society*. Retrieved from: https://www.psychology.org.au/About-Us/What-we-do/advocacy/Advocacy-social-issues/Environment-climate-change-psychology/Resources-for-Psychologists-and-others-advocating/The-psychology-of-climate-change-denial

BBC (2019). School strike for climate: Protests staged around the world. Retrieved from: https://www.bbc.com/news/world-48392551

Berkeley Earth (2020). Global Temperature Report for 2018. Retrieved from: http://berkeleyearth.org/2018-temperatures/

BGA (n.d.). What is Geophysics? British Geographical Association. Retrieved from: https://geophysics.org.uk/what-is-geophysics/

Blinken, Anthony, J. (2021). The United States Officially Rejoins the Paris Agreement. U.S Department of State. Retrieved from: https://www.state.gov/the-united-states-officially-rejoins-the-paris-agreement/

MORTEN BUTTLER/BLOOMBERG, MORTEN (2021). Greenland Bans All Future Oil Exploration Citing Climate Concerns. *Time*. Retrieved from: https://time.com/6080933/greenland-bans-oil-exploration/

Campbell, J. E., Dettrey, B. J. and Yin, H. (2010). The Theory of Conditional Retrospective Voting: Does the Presidential Record Matter Less in Open-Seat Elections? *The Journal of Politics , Vol. 72, No. 4 (Oct., 2010), pp. 1083-1095.* Retrieved from: https://www.jstor.org/stable/10.1017/s002238161000054x;ttps://www.jstor.org/stable/pdf/10.1017/s002238161000054x.pdf?refreqid=excelsior%3 A361531f-aaab49a5913857647100 27d4b

Catalina (2020). Industrialization and Climate Change. Retrieved from: https://earth.usc.edu/~stott/Catalina/Industrial.html

CDC. CERC: Psychology of a Crisis. CDC Emergency Preparedness. Retrieved from: https://emergency.cdc.gov/cerc/ppt/CERC_Psychology_of_a_Crisis.pdf

CDP (2017). New report shows just 100 companies are source of over 70% of emissions. Retrieved from: https://www.cdp.net/en/articles/

media/new-report-shows-just-100- companies-are-source-of-over-70-of-emissions

Charner, F. and Kottasová, I. (2020). Brazil's Bolsonaro rejects Biden's offer of $20 billion to protect the Amazon. *CNN.* Retrieved from: https://www.cnn.com/2020/09/30/americas/brazil-bolsonaro-biden-amazon-intl/index.html

Chebreyesus, T. A. (2019). Climate Change Is Already Killing Us: How Our Warmer and Wetter Planet Is Getting Sicker and Deadlier by the Day. Foreign Affairs March/April, 2020. Retrieved from: https://www.foreignaffairs.com/articles/2019-09-23/climate-change- already-killing-us?utm_medium=cpc&utm_source=google&utm_campaign=cfr_grant_lists&gclid=EAIa IqobChMIv-bU5beQ6AIVT_DACh21hgLEEAAYAyAAEgIUHvD_BwE

Chemnick, J., and Frank, T. (2020). Climate Change Once Again Left Out of Trump's Federal Budget. Retrieved from: https://www.scientificamerican.com/article/climate-change-once-again-left-out-of-trumps-federal-budget/

Cool earth (2018). IPCC Global Warming Special Report 2018 | What does it actually mean? Retrieved from: https://www.coolearth.org/2018/10/ipcc-report-2/?gclid=EAIaIQobChMIhYTjq8OX-6AIVhobACh2TDAG-EAAYASAAEgKg0fD_BwE

Culatta, R. (2020). Connectionism (Edward Thorndike). Instructional Design. Retrieved from: https://www.instructionaldesign.org/theories/connectionism/

CRED (2018). Natural Disaster (2018). Center for Research on the Epidemiology of natural disasters. Retrieved from: https://www.cred.be/natural-disasters-2018

CUSHMAN, J. H. and HIRJI, Z (2017). Paris Agreement: Trump's Climate Exit Risks U.S. Economy, World's Welfare. Inside Climate News.

Retrieved from: https://insideclimatenews.org/news/01062017/donald-trump-paris-climate-change- agreement- decided?gclid=EAIaIQobChMIvPbV6p_96wIVrx6tBh088Q4DEAAYBCAAEgJl9fDB_ wE

De Groot JIM, Steg L. 2010. Relationships between value orientations, self-determined motivational types and pro-environmental behavioural intentions. *J. Environ. Psychol. 30:368–78*

Denchak, M. (2017). Global Climate Change: What You Need to Know. Natural Resources Defense Council (NRDC). Retrieved from: https://www.nrdc.org/stories/global-climate-change-what-you-need-know?gclid=EAIaIQobChMIv- bU5beQ6A-IVT_DACh21hgLEEAAYAiAAEgIxIvD_BwE

Denechak, Mellisa (2018). Fossil Fuels: The Dirty Facts. Natural Resources Defense Council Retrieved from: https://www.nrdc.org/stories/fossil-fuels-dirty-facts

Denchak, M. (2018). Paris Climate Agreement: Everything You Need to Know. Natural Resources Defense Council (NRDC). Retrieved from: Retrieved from: https://www.nrdc.org/stories/paris-climate-agreement-everything-you-need-know

Dennis, Brady and Joselow, Maxine (2022). U.S. climate promises hang in the balance as Manchin upends talks. The Washington Post. Retrieved from: https://www.washingtonpost.com/climate-environment/2022/07/15/manchin-climate-biden-paris-agreement/

Deutsche Welle (2020). Psychology behind climate inaction: How to beat the 'doom barrier' Retrieved from: https://www.dw.com/en/psychology-behind-climate-inaction-how-to-beat-the-doom-barrier/a-48730230

ECU (2019). What are the Causes of Water Pollution? Water Resource Policy and Management. East Central University. Retrieved from: https://online.ecok.edu/articles/causes-of-water- pollution/

ENERSYSTEMS, INC. (2023). Retrieved from: https://opencorporates.com/companies/us_wv/123548

Environmental Pollution Centers (2022). Retrieved from: https://www.environmentalpollutioncenters.org/water/causes/

EPA (2017). Causes of Climate Change. *United States Environmental Protection Agency.* Retrieved from: https://www.youtube.com/watch?v=1Yp91zKBGgY

EPA (2022). Drinking Water: What are the trends in the quality of drinking water and their effects on human health? Environmental Protection Agency (EPA). Retrieved from: https://www.epa.gov/report-environment/drinking-water

Exploratorium (2020). Keeping an Eye on our changing planet. Retrieved from: https://www.exploratorium.edu/climate?gclid=EAIaIQobChMIg6rvnrmO5wIVDNvACh1Iew5fEAMYASAAEgJnlPD_BwE

Feinberg M, Willer R. 2013. The moral roots of environmental attitudes. *Psychol. Sci. 24:56–62*

Fleischer, N. L. (2014). Outdoor air pollution, preterm birth, and low birth weight: Analysis of the World Health Organization global survey on maternal and perinatal health. *Environmental Health Perspectives.* Vol. 123, April 2014, p. 425. Doi: 10.1289/ehp.1306837.

Fraser, R. (2010). Why Have Natural Disasters Increased? There is a grave warning message in the worldwide escalation of natural disasters over the years. Retrieved from: http://realneo.us/content/why-have-natural-disasters-increased-there-grave-warning-message-worldwide-escalation-natura

Funk, C. and Hefferon, M. (2019). U.S. Public views on climate and energy. Pew Research Center. Retrieved from: https://www.pewresearch.org/science/2016/10/04/public-views-on-climate-change-and-climate-scientists/

GAO (2014). Government Accountability Office on Climate Change Spending. *U.S Government Accountability Office.* Retrieved from: https://www.gao.gov/key_issues/climate_change_funding_management/issue_summary

Gleick, Peter (2017). Statements on Climate Change from Major Scientific Academies, Societies, and Associations (January 2017 update). Retrieved from: https://scienceblogs.com/significantfigures/index.php/2017/01/07/statements-on-climate- change-from-major-scientific-academies-societies-and-associations-january-2017-update

Globalchange.gov (2018). Fourth National Climate Assessment. U.S Global Change Research Program. Retrieved from: https://www.globalchange.gov/nca4

Gorman, S. Ph.D., and Gorman, J. M. MD (2019). Climate Change Denial Facing a reality too big to believe. Psychology Today. Retrieved from: https://www.psychologytoday.com/us/blog/denying-the-grave/201901/climate-change- denial

Gray, P. (2011). *Psychology* (6th ed.) New York: Worth Publishers.

GreenPeace (2022). Koch Industries Pollution. Retrieved from: https://www.greenpeace.org/usa/fighting-climate-chaos/climate-deniers/koch-industries/koch-industries-pollution/

Grossman, E. (2016). How fossil fuel use threatens kids' health. *Science News for Students.* Retrieved from: https://www.sciencenewsforstudents.org/article/how-fossil-fuel-use- threatens-kids-health

Hausfather, Dr. Zeke and Friedlingstein, Prof Pierre (2022). Analysis: Global CO2 emissions from fossil fuels hit record high in 2022. Retrieved from: https://www.carbonbrief.org/analysis-global-co2-emissions-from-fossil-fuels-hit-record- high-in- 2022/#:~:text=Global%20 carbon%20dioxide%20emissions%20from,by%20the%20Glob al%20 Carbon%20Project.

Heimlich, Joe E. & Ardoin, Nicole M. (2008) Understanding behavior to understand behavior change: a literature review, Environmental Education Research, 14:3, 215-237, DOI: 10.1080/13504620802148881. Retrieved from: https://www.tandfonline.com/doi/full/10.1080/13504620802148881

HERSHER, REBECCA (2019). U.S. Formally Begins To Leave The Paris Climate Agreement. Retrieved from: https://www.npr.org/2019/11/04/773474657/u-s-formally-begins-to-leave-the-paris-climate-agreement

IEA (2020). Oil market report. *International Energy Agency*. Retrieved from: https://www.iea.org/topics/oil-market-report

Inside Climate News (2017). Global Climate Treaty Retrieved from: https://insideclimatenews.org/topics/global-climate-treaty

Instructional Design.org (n.d). Connectionism (Edward Thorndike). Retrieved from: https://www.instructionaldesign.org/theories/connectionism/

IPCC (2014). Climate Change 2014 Synthesis Report. *INTERGOVERNMENTAL PANEL ON climate change (IPCC)*. Retrieved from: https://www.ipcc.ch/site/assets/uploads/2018/05/SYR_AR5_FINAL_full_wcover.pdf

IPCC (2018). Special Report: Global Warming of 1.5 degrees Centigrade. *Intergovernmental Panel on Climate Change*. Retrieved from: https://www.ipcc.ch/sr15/

IPCC (2022). IPCC Press release. *Intergovernmental Panel on Climate Change*. Retrieved from: https://www.ipcc.ch/report/ar6/wg2/resources/press/press-release/ ; https://www.ipcc.ch/report/ar6/wg2/downloads/press/IPCC_AR6_WGII_PressRelease-English.pdf

Keys, John Maynard (1936). *The General Theory of Employment, Interest and Money.* Harcourt, Brace and Company, and printed in the U.S.A. by the Polygraphic Company of America, New York; First Published: Macmillan Cambridge University Press, for Royal Economic Society in 1936. Retrieved from: https://www.marxists.org/reference/subject/economics/keynes/general-theory/index.htm

Chapter 2. The Postulates of Classical Economics. Retrieved from: https://www.marxists.org/reference/subject/economics/keynes/general- theory/ch02.htm#vi

King, M., W. (2019). Cognitive biases that ensured our initial survival now make it difficult to address long-term challenges that threaten our existence, like climate change. But they can help us too. Retrieved from: https://www.bbc.com/future/article/20200109-is-it-wrong-to-be-hopeful-about-climate-change

Knutson, T.; Kossin, J. P.; Mears, C.; Perlwitz, J.; Wehner, M. F. (2017). "Chapter 3: Detection and Attribution of Climate Change" (PDF). *In USGCRP2017.* Retrieved from: https://science2017.globalchange.gov/downloads/CSSR_Ch3_Detection_and_Attribution .pdf

Kober, A. (2017). WITHDRAWAL FROM PARIS CLIMATE AGREEMENT: STATEMENT FROM AMERICAN RIVERS RETRIRVED FROM: HTTPS://WWW.AMERICANRIVERS.ORG/CONSERVATION-RESOURCE/WITHDRAWAL-PARIS-CLIMATE-AGREEMENT-STATEMENT-AMERICAN- RIVERS/?GCLID=EAIAIQOBCHMIPFKZN8OQ6AIVJIBACH3NZWMQEAAYBCAAEGIG3VD BWE

Kotcher, J., Maibach, E. & Choi, W. (2019)/ Fossil fuels are harming our brains: identifying key messages about the health effects of air pollution from fossil fuels. *BMC Public Health* **19,** 1079 (2019). https://doi.org/10.1186/s12889-019-7373-1. Retrieved from: https://bmcpublichealth.biomedcentral.com/articles/10.1186/s12889-019-7373-1#citeas

Lavelle, M. (2020). A Climate Change Skeptic, Mike Pence Brought to the Vice Presidency Deep Ties to the Koch Brothers. *Inside Climate News.* Retrieved from: https://insideclimatenews.org/news/31082020/candidate-profile-mike-pence-climate- change-election-2020

LeVine, M. and Ferris, S. (2020). McConnel says coronavirus relief package 'unlikely' before election. *POLITICO.* Retrieved from: https://www.politico.com/news/2020/10/09/mcconnell-coronavirus-relief-package 428315

Lewis, Michelle (2022). The real reason Joe Manchin is sabotaging the US clean energy plan [update]. Retrieved from: https://electrek.co/2022/07/14/joe-manchin-sabotaging-us- clean-energy-plan/

LINDSEY, REBECCA & DAHLMAN, LUANN (2023). REVIEWED BY BLUNDEN, JESSICA. Climate Change: Global Temperature. Climate.gov., Department of Commerce. NORA. Retrieved from: https://www.climate.gov/news-features/understanding-climate/climate-change-global-temperature#:~:text=According%20to%20NOAA's%202021%20Annual,0.18%20%C2% B0C)%20per%20decade

Lubell, Mark, Zahran, Sammy, Vedlitz, Arnold (2007). Collective Action and Citizen Responses to Global Warming. Polit Behav (2007) 29:391–413 DOI 10.1007/s11109-006-9025-2. Retrieved from: https://link.springer.com/content/pdf/10.1007/s11109-006-9025-2.pdf https://link.springer.com/article/10.1007/s11109-006-9025-2

Lyons, B., Hasell, A. and Stroud, N. J. (2018). Extreme Weather and Climate Skeptics Are Research Focus of APPC Postdocs. *Annenberg Public Policy Center* of the University of Pennsylvania. Retrieved from: https://www.annenbergpublicpolicycenter.org/climate-skeptics-extreme-weather/

Mai, H. J. (2021). U.S. Officially Rejoins Paris Agreement On Climate Change. NPR. Retrieved from: https://www.npr.org/2021/02/19/969387323/u-s-officially-rejoins-paris-agreement- on-climate-change

Mantle314 (2015). The Paris Agreement is a Universal Call to Action and a Market Signal to Heed. Retrieved from: https://mantle314.com/insights/the-paris-agreement-is-a-universal-call-to-action-and-a-market-signal-to-heed/

McKENZIE, R. B. (2019). The Climate-Change Doomsday Trap. Retrieved from: https://www.cato.org/sites/cato.org/files/serials/files/regulation/2019/6/reg-v42n2-2.pdf

McKibben, B. (2019). Money Is the Oxygen on Which the Fire of Global Warming Burns. The New Yorker. Retrieved from: https://www.newyorker.com/news/daily-comment/money-is-the-oxygen-on-which-the-fire-of-global-warming-burns

Melillo, Jerry M., Terese (T.C.) Richmond, and Gary W. Yohe, Eds. (2014). Climate Change Impacts in the United States: *The Third National Climate Assessment. U.S. Global Change Research Program, 841 pp*. doi:10.7930/J0Z31WJ2.

Mgbemene, C. A., Nnaji, C. C. and Nwozor, C. (2016). Industrialization and its Backlash: Focus on Climate Change and its Consequences. *Journal of Environmental Science and Technology. Volume 9 (4): 301-316, 2016.* Retrieved from:" https://scialert.net/fulltextmobile/?-doi=jest.2016.301.316

Miles-Novelo, A., Anderson, C.A. (2019). Climate Change and Psychology: Effects of Rapid Global Warming on Violence and Aggression. *Curr Clim Change Rep* **5,** 36–46 (2019). https://doi.org/10.1007/

s40641-019-00121-2. Retrieved from: https://link.springer.com/content/pdf/10.1007/s40641-019-00121-2.pdf ; https://link.springer.com/article/10.1007/s40641-019-00121-2

MITRA, M. N. (2018). The US Climate Change Report Trump Didn't Want You to Know About. *Earth Island Journal* Retrieved from: https://www.earthisland.org/journal/index.php/articles/entry/the-us-climate-change-report-trump-€-want-you-to-know-about?gclid=EAIaIQobChMIpbbqu4P96wIVOx- tBh34iQA1EAAYBCAAEgKh_PD_BwE

Mooney, C and Freedman, A. (2019). The world needs a massive carbon tax in just 10 years to limit climate change, IMF says. The Washington Post. Retrieved from: https://www.washingtonpost.com/climate-environment/2019/10/10/world-needs-massive- carbon-tax-just-years-limit-climate-change-imf-says/

Mooney, Chris and Stevens, Harry (2022). The U.S. plan to avoid extreme climate change is running out of time. *The Washington Post.* Retrieved from: https://www.washingtonpost.com/climate-environment/2022/07/18/climate-change- manchin-math/?utm_campaign=wp_post_most&utm_medium=email&utm_source=newsletter&psrc=nl_most&carta-url=https%3A%2F%2Fs2.washingtonpost.com%2Fcar- lntr%2F3768503%2F62d58160c-fe8a21601f559b0%2F626130676f02666fddec8e6e%2F8 %2F72%2F-62d58160cfe8a21601f559b0&wp_cu=039fec53babffcac0026a1b-96420fbba% 7CDD28533A268D5D95E0530100007F33E6

Mortillaro, N. (2018). The psychology of climate change: Why people deny the evidence. Retrieved from: https://www.cbc.ca/news/technology/climate-change-psychology- 1.4920872

Mullins, Jamie T. and White, Corey (2019). Temperature and mental health: Evidence from the spectrum of mental health outcomes. *Journal of Health Economics, Volume 68, 2019, 102240, ISSN 0167-6296*, https://

doi.org/10.1016/j.jhealeco.2019.102240. Retrieved from: https://www.sciencedirect.com/science/article/pii/S016762961830105X

Nardinelli, Clark. Industrial Revolution and the Standard of Living. *U.S. Food and Drug Administration.* Retrieved from: https://www.econlib.org/library/Enc/IndustrialRevolutionandtheStandardofLiving.html

NASA (2020). Global Temperature. *NASA Global Climate Change.* Retrieved from: https://climate.nasa.gov/vital-signs/global-temperature/

NASA (2020A). Temperature Anomalies over Land and over Ocean.

Retrieved from: https://data.giss.nasa.gov/gistemp/graphs_v4/

NASA (2020B). Global Climate Change: Vital signs of the Planet. Retrieved from: https://climate.nasa.gov/solutions/adaptation-mitigation/

Nasa (2021). Global Temperature. National Aeronautics and Space Association. Retrieved from: https://climate.nasa.gov/vital-signs/global-temperature/

Hash, Matthew (2022). These 36 World Cities Will Be Underwater

First. Retrieved from: https://theswiftest.com/underwater-cities/

Nazir, F. B. (2018). Stimulus-Response Theory: A Case Study in the Teaching and Learning of Malay Language Among Year 1 Pupils. *The Journal of Social Sciences Research, ISSN€: 2411-9458, ISSN(p): 2413-6670, Vol. 4, Issue. 10, pp: 153-157,* 2018 URL: https://arpgweb.com/journal/journal/7 DOI:https://doi.org/10.32861/journal.7.2018.410.153.157

Nerlich, Brigitte, Koteyko, Nelya, and Brown, Brian (2010). Theory and language of climate change communication. John Wiley & Son s,

Ltd., Volume 1, January/February 2010. Retrieved from: https://wires.onlinelibrary.wiley.com/doi/pdf/10.1002/wcc.2

Newell, B. R., McDonald, R. I., Brewer, M. and Hayes, B. K. (2017). The Psychology of Environmental Decisions. *Annu. Rev. Environ. Resour. 2014. 39:443–67 The Annual Review of Environment and Resources*, doi: 10.1146/annurev-environ010713094623. Retrieved from: http://www.columbia.edu/~rim2114/publications/AnnRev-2014-Newell- McDonald-Brewer-Hayes.pdf

Nielsen, K. S., Clayton, S., Stern, P., C., Dietz, T., Capstick S. and Whitmars, L.(2020). How Psychology Can Help Limit Climate Change. *Researchgate.* Retrieved from: https://www.researchgate.net/publication/339146241_How_Psychology_Can_Help_Limit_Climate_Change

North, E., Halden, R. U. (2013). Plastics and Environmental Health: The Road Ahead. RevEnviron Health. 2013; 28(1): 1–8., doi: 10.1515/reveh-2012-0030. Retrieved from: https://www.ncbi.nlm.nih.gov/pmc/articles/PMC3791860/

Nova Biologicals Team (2018). TOP 6 CAUSES OF WATER POLLUTION AND HOW TO REDUCE THE RISKS. NOVA BIOLOGICALS. RETRIEVED FROM: HTTPS://WWW.NOVATX.COM/DRINKING-WATER/TOP-6-CAUSES-WATER-POLLUTION-REDUCE-RISKS/

Ortiz, Diego Arguedas (2020). Is it wrong to be hopeful about climate change? Retrieved from: https://www.bbc.com/future/article/20200109-is-it-wrong-to-be-hopeful-about-climate-change

Oxford Reference (2020). Thorndike's stimulus–Response theory of learning. Retrieved from: https://www.oxfordreference.com/view/10.1093/oi/authority.20110803104425878

Oxford Reference (2020). Law of exercise. Retrieved from: https://www.oxfordreference.com/view/10.1093/oi/authority.20110803100054849

Oxford Reference (2020). Law of Readiness. Retrieved from: https://www.oxfordreference.com/view/10.1093/oi/authority.20110803100054989

Perera, Frederica (2016). Multiple threats to child health from fossil fuel combustion: Impacts of air pollution and climate change. *Environmental Health Perspectives.* Published early online June 21, 2016. Do1:10.1289/EHP299.

Remnick, D. (2016, November 28). It happened here. The New Yorker, p. 92. Google Scholar

Ritchie, H. and Roser, M. (2014). Natural Disasters. Our World in Data, University of Oxford. Retrieved from: https://ourworldindata.org/natural-disasters

ROWELL, A. (2017). Climate Change is Costing Billions & Set to Get Worse, Says GAO. Retrieved from: http://priceofoil.org/2017/10/25/climate-change-is-costing-billions-set- to-get-worse-says-gao/

Schreck III, C. J., K. R. Knapp, and J. P. Kossin (2014). The Impact of Best Track Discrepancies on Global Tropical Cyclone Climatologies using IBTrACS. National Centers for Environmental Information. *Monthly Weather Review*, 142, 3881 3899. doi:10.1175/MWR-D-14-00021.1. Retrieved from: https://www.ncdc.noaa.gov/ibtracs/index.php?name=-climatology

Schouten, Fredreka (2022). Joe Manchin, who just torpedoed Democrats' climate agenda, has long ties to coal industry. CNN Politics. Retrieved from: https://www.cnn.com/2022/07/15/politics/joe-manchin-coal-financial-interests- climate/index.html

Semenza, Jan, C., PhD, MPH, David E. Hall, PhD, Daniel J. Wilson, Brian D. Bontempo, PhD, David J. Sailor, PhD, Linda A. George, PhD

(2008). Public Perception of Climate Change: Voluntary Mitigation and Barriers to Behavior Change. *American Journal of Preventive Medicine, Volume 35, Issue 5, November 2008, Pages 479-487. Published by Elsevier Inc. doi:10.1016/j.amepre.2008.08.020.* Retrieved from: https://www.sciencedirect.com/science/article/pii/S0749379708006831

Sina (2010). Study Shows dramatic rise in natural disasters over the past decade. Retrieved from: http://english.sina.com/technology/p/2010/0128/302222.html

Study.com (2020). Connectionism: Overview & Practical Teaching Examples Chapter 4 / Lesson 11 Transcript. *Study.com. Education 101: Foundations of Education/Social Science Courses.* Retrieved from: https://study.com/academy/lesson/connecionism-overview-practical-teaching examples.html

Tilly, **Dr. Richard H. (2018).** Industrialization as an Historical Process. Retrieved from: https://brewminate.com/industrialization-as-an-historical-process/

Timperley, J. (2020). Who is really to blame for climate change? Retrieved from: https://www.bbc.com/future/article/20200618-climate-change-who-is-to-blame-and-why-does-it-matterhttps://www.bbc.com/future/article/20200618-climate-change-who-is-to-blame-and-why-does-it-matter

Truelove, Heather Barnes and Parks, Craig (2012). Perceptions of behaviors that cause and mitigate global warming and intentions to perform these behaviors. *Journal of Environmental Psychology 32 (2012) 246e259.* Retrieved from: https://www.sciencedirect.com/science/article/pii/S0272494412000254

Union of Concerned Scientist (2018). Cars, Trucks, Buses and Air Pollution. Retrieved from: https://www.ucsusa.org/resources/cars-trucks-buses-and-air-pollution

United Nations (2019). Climate Change. Retrieved from: https://www.un.org/en/sections/issues- depth/climate-change/

Van Boven, L., Ehret, P. J., & Sherman, D. K. (2018). Psychological Barriers to Bipartisan Public Support for Climate Policy. *Perspectives on Psychological Science, 13*(4), 492 507. https://doi.org/10.1177/1745691617748966. Retrieved from: https://journals.sagepub.com/doi/10.1177/1745691617748966#articleCitationDownloadContainer

Victor, D. G., Obradovich, N. and Amaya, D. J. (2017). Why the wiring of our brains makes it hard to stop climate change. *Brookings Institute.* Retrieved from: https://www.brookings.edu/blog/planet-policy/2017/09/18/why-the-wiring-of-our-brains- makes-it-hard-to-stop-climate-change/

White House (2017). Statement by President Trump on the Paris Climate Accord. Retrieved from: https://www.whitehouse.gov/briefings-statements/statement-president-trump-paris- climate-accord/

Whitehouse, Senators Sheldon (D-RI) and Schumer, Chuck (D-NY) (2019). CLIMATE CHANGE AND DARK MONEY. RETRIEVED FROM: HTTPS://WWW.WHITEHOUSE.SENATE.GOV/NEWS/OP-EDS/CLIMATE-CHANGE-AND-DARK-MONEY

WHO (2020). Air Pollution. *World Health Organization.* Retrieved from: https://www.who.int/health-topics/air-pollution#tab=tab_1

Why is plastic harmful? Plastics Pollution Coalition. Retrieved from:https://plasticpollutioncoalition.zendesk.com/hc/en-us/articles/222813127-Why-is- plastic harmful

Williamson, K., Satre-Meloy, A., Velasco, K., & Green, K., 2018. Climate Change Needs Behavior Change: Making the Case for Behavioral Solutions to Reduce Global Warming. Arlington, VA: Ra. *The Center for Behavior & the Environment at Rare.* Retrieved from: https://rare.org/wp-content/uploads/2019/02/2018-CCNBC-Report.pdf

Wilczek, Frank (2013). Einstein's Parable of Quantum Insanity. Quanta Magazine on September 23, 2015. Retrieved from: https://www.scientificamerican.com/article/einstein-s-parable- of-quantum-insanity/

Woods, Tyler (2022). What Rising Temperatures Mean for Our Mental Health. Psychology Today. Retrieved from: https://www.psychologytoday.com/us/blog/evidence-based- living/202209/what-rising-temperatures-mean-our-mental-health

Working Group, 2022. Climate Change 2022: Impact, Adaptation, Vulnerability. Intergovernmental Panel on Climate Change (IPCC). Retrieved from: https://report.ipcc.ch/ar6wg2/pdf/IPCC_AR6_WGII_SummaryForPolicymakers.pdf https://www.ipcc.ch/report/ar6/wg2/about/frequently-asked-questions/keyfaq1 https://www.theguardian.com/environment/2022/feb/28/what-at-stake-climate-crisis- report-everything https://report.ipcc.ch/ar6wg2/pdf/IPCC_AR6_WGII_IntroductionWGII.pdf

World Bank. 2010. World Development Report 2010: Development and Climate Change. Washington, DC. © *World Bank. https://openknowledge.worldbank.org/handle/10986/4387 License: CC BY 3.0 IGO.* Retrieved from: https://openknowledge.worldbank.org/handle/10986/4387

Worldometer (2020). World Population by Year. **Retrieved from:** https://www.worldometers.info/world-population/world-population-by-year/

WorldOMeter (2023). World Population: Past, Present, and Future. Retrieved from: https://www.worldometers.info/world-population/

World Population Prospects (2019 Revision) – United Nations population estimates and projections; Historical Estimates of World Population. 2020 World Population by Country. Retrieved from: http://worldpopulationreview.com/

Xiaoqi Zheng, Yonglong Lu, Jingjing Yuan, Yvette Baninla, Sheng Zhang, Nils Chr. Stenseth, Dag O. Hessen, Hanqin Tian, Michael Obersteiner, and Deliang Chen (2020). Drivers of change in China's energy-related CO_2 **emissions** *PNAS January 7, 2020 117 (1) 29-36; first published December 23, 2019.* https://doi.org/10.1073/pnas.1908513117. Retrieved from: https://www.pnas.org/content/117/1/29

Yifu Lin, Justin and Rosenblatt, David (2013). Shifting Patterns of Economic Growth and Rethinking Development. *World Bank · Chief Economist's Office.* Retrieved from: https://www.researchgate.net/figure/Development-since-the-industrial-revolution-Per- capita-GDP-measured-in-1990_fig8_254072591

APPENDIX 1

Statements on Climate Change from Major Scientific Academies, Societies, and Associations (January 2017 update). The following credible and expert entities support actions to control climate change and reduce global warming. What is the problem with humans that they, in their ignorance, disagree with experts, who know what they are doing?

Science has been monitoring and tracking the role of humans in influencing and altering the global climate for over a century. Science has gained an expert understanding of the primary cause of global warming, humans.

"There is no longer any reasonable doubt that humans are altering the climate, that those changes will grow in scope and severity in the future, and that the economic, ecological, and human health consequences will be severe" (Gleick, 2017).

The number and scope of these statements acknowledging climate change and the need to mitigate its impact is impressive. Not a single major scientific organization or national academy of science on earth denies that the climate is changing, that humans are responsible, and that some form of action should be taken to address the risks to people and the planet. This is a primary reason that when politicians deny climate change, it is proof of their inability, lack of intelligence on the subject and indication they are not qualified to serve the American or International community in a position of leadership.

"The false argument has a name: the ***Galileo Gambit***. It is used by those who deny well-established scientific principles such as the theory of climate change as follows: Because Galileo was mocked and criticized for his views by a majority, but later shown to be right, current minority views that are mocked and criticized must also be right. The obvious flaw in the Galileo Gambit is that being criticized for one's views does not correlate with being right – especially when the criticism is based on scientific evidence. Galileo was right because ***the scientific evidence supported him***, not because ***he was mocked and criticized***. The late professor Carl Sagan addressed this use of the Galileo Gambit in a humorous way when he noted:"

> *"But the fact that some geniuses were laughed at does not imply that all who are laughed at are geniuses. They laughed at Columbus, they laughed at Fulton, they laughed at the Wright Brothers. But they also laughed at Bozo the Clown" (Broca's Brain, 1979).*

These statements and declarations about climate change by the world's leading scientific organizations represent the most expert summary of the state of knowledge and concern about the environmental conditions of the world. These statements from the world's leading Scientific Academies, Societies, and Associations provide the basis and rationale for people throughout the world, voters, politicians and industrialist to take climate change seriously. "The world ignores them at its peril."

The following is a credible world-wide list of scientific, engineering, and health organizations that have issued statements about human-caused climate change (as of January 2017).

Allergy and Asthma Network April 2016

The AAN is a signatory to the April 2016 statement: http://www.lung.org/our-initiatives/healthy-air/outdoor/climate-change/declaration-on-climate-change.html?referrer=https://www.google.com/

American Academy of Pediatrics November 2015

http://pediatrics.aappublications.org/content/136/5/992

Rising global temperatures are causing major physical, chemical, and ecological changes in the planet. There is wide consensus among scientific organizations and climatologists that these broad effects, known as "climate change," are the result of contemporary human activity. Climate change poses threats to human health, safety, and security, and children are uniquely vulnerable to these threats...

American Association for the Advancement of Science (AAAS)
December 9, 2006, reaffirmed December 2009

http://www.aaas.org/news/aaas-reaffirms-statements-climate-change-and-integrity

The scientific evidence is clear: global climate change caused by human activities is occurring now, and it is a growing threat to society. Accumulating data from across the globe reveal a wide array of effects: rapidly melting glaciers, destabilization of major ice sheets, increases in extreme weather, rising sea level, shifts in species ranges, and more. The pace of change and the evidence of harm have increased markedly over the last five years. The time to control greenhouse gas emissions is now.

[The AAAS has also signed onto more recent letters on climate from an array of scientific organizations, including the June 28, 2016 letter to the U.S. Congress: https://www.eurekalert.org/images/2016climate-letter6-28-16.pdf]

American Association of Wildlife Veterinarians October 2008

http://www.aawv.net/AAWVPositionClimateChangeFinal.doc

There is widespread scientific agreement that the world's climate is changing and that the weight of evidence demonstrates that anthropogenic factors have and will continue to contribute significantly to global warming and climate change. It is anticipated that continuing changes to the climate will have serious negative impacts on public, animal and ecosystem health due to extreme weather events, changing disease transmission dynamics, emerging and re-emerging diseases, and alterations to habitat and ecological systems that are essential to wildlife conservation. Furthermore, there is increasing recognition of the inter-relationships of human, domestic animal, wildlife, and ecosystem health as illustrated by the fact the majority of recent emerging diseases have a wildlife origin. Consequently, there is a critical need to improve capacity to identify, prevent, and respond to climate-related threats. The following statements present the American Association of Wildlife Veterinarians (AAWV) position on climate change, wildlife diseases, and wildlife health….

American Astronomical Society June 2, 2004, Endorsement of AGU Statement on Climate Change

The American Geophysical Union (AGU) notes that human impacts on the climate system include increasing concentrations of greenhouse gases in the atmosphere, which is significantly contributing to the warming of the global climate. The climate system is complex, however, making it difficult to predict detailed outcomes of human-induced change: there is as yet no definitive theory for translating greenhouse gas emissions into forecasts of regional weather, hydrology, or response of the biosphere. As the AGU points out, our ability to predict global climate change, and to forecast its regional impacts, depends directly on improved models and observations.

The American Astronomical Society (AAS) joins the AGU in calling for peer-reviewed climate research to inform climate-related policy decisions, and, as well, to provide a basis for mitigating the harmful effects of global change and to help communities adapt and become resilient to extreme climatic events.

In endorsing the "Human Impacts on Climate" statement, the AAS recognizes the collective expertise of the AGU in scientific subfields central to assessing and understanding global change, and acknowledges the strength of agreement among our AGU colleagues that the global climate is changing and human activities are contributing to that change.

American Chemical Society Policy Statement 2013-2016

https://www.acs.org/content/acs/en/policy/publicpolicies/promote/globalclimatechange.html

Careful and comprehensive scientific assessments have clearly demonstrated that the Earth's climate system is changing in response to growing atmospheric burdens of greenhouse gases (GHGs) and absorbing aerosol particles. (IPCC, 2007) Climate change is occurring, is caused largely by human activities, and poses significant risks for—and in many cases is already affecting—a broad range of human and natural systems. (NRC, 2010a) The potential threats are serious, and actions are required to mitigate climate change risks and to adapt to deleterious climate change impacts that probably cannot be avoided. (NRC, 2010b, c)

This statement reviews key probable climate change impacts and recommends actions required to mitigate or adapt to current and anticipated consequences.

Climate Change Impacts

...comprehensive scientific assessments of our current and potential future climates clearly indicate that climate change is real, largely attributable to emissions from human activities, and potentially a very serious problem. This sober conclusion has been recently reconfirmed by an in-depth set of studies focused on "America's Climate Choices" (ACC) conducted by the U.S. National Academies (NRC, 2010a, b, c, d). The ACC studies, performed by independent and highly respected teams of scientists, engineers, and other skilled professionals, reached the same general conclusions that were published in the latest comprehensive assessment conducted by the International Panel on Climate Change (IPCC, 2007)...

The range of observed and potential climate change impacts identified by the ACC assessment include a warmer climate with more extreme weather events, significant sea level rise, more constrained fresh water sources, deterioration or loss of key land and marine ecosystems, and reduced food resources— many of which may pose serious public health threats. (NRC, 2010a) The effects of an unmitigated rate of climate change on key Earth system components, ecological systems, and human society over the next 50 years are likely to be severe and possibly irreversible on century time scales...

[The ACS has also signed onto more recent letters on climate from an array of scientific organizations, including the June 28, 2016 letter to the U.S. Congress: https://www.eurekalert.org/images/2016climate-letter6-28-16.pdf]

American College of Preventive Medicine February 24, 2006, Policy Number 2006-002C

http://web.archive.org/web/20060925182111/http://www.acpm.org/2006-002(C).htm

Climate Change-Abrupt Climate Change and Public Health Implications

BE IT RESOLVED,

THAT: The American College of Preventive Medicine (ACPM) accept the position that global warming and climate change is occurring, that there is potential for abrupt climate change, and that human practices that increase greenhouse gases exacerbate the problem, and that the public health consequences may be severe.

THAT: The ACPM staff and appropriate committees continue to explore opportunities to address this matter, including sessions at Preventive Medicine conferences and the development of a policy position statement as well as other modes of communicating this issue to the ACPM membership.

[The ACPM is also a signatory to the April 2016 statement: http://www.lung.org/our-initiatives/healthy-air/outdoor/climate-change/declaration-on-climate-change.html?referrer=https://www.google.com/]

American Geophysical Union

Adopted by the American Geophysical Union December 2003; Revised and Reaffirmed December 2007, February 2012, August 2013.

http://sciencepolicy.agu.org/files/2013/07/AGU-Climate-Change-Position-Statement_August-2013.pdf

Human-Induced Climate Change Requires Urgent Action

Humanity is the major influence on the global climate change observed over the past 50 years. Rapid societal responses can significantly lessen negative outcomes. Human activities are changing Earth's climate. At the global level, atmospheric concentrations of carbon dioxide and other heat-trapping greenhouse gases have increased sharply since the Industrial Revolution. Fossil fuel burning dominates this increase.

Human-caused increases in greenhouse gases are responsible for most of the observed global average surface warming of roughly 0.8°C (1.5°F) over the past 140 years. Because natural processes cannot quickly remove some of these gases (notably carbon dioxide) from the atmosphere, our past, present, and future emissions will influence the climate system for millennia.

Extensive, independent observations confirm the reality of global warming. These observations show large-scale increases in air and sea temperatures, sea level, and atmospheric water vapor; they document decreases in the extent of mountain glaciers, snow cover, permafrost, and Arctic sea ice. These changes are broadly consistent with long understood physics and predictions of how the climate system is expected to respond to human-caused increases in greenhouse gases. The changes are inconsistent with explanations of climate change that rely on known natural influences...

[The AGU has also signed onto more recent letters on climate from an array of scientific organizations, including the June 28, 2016 letter to the U.S. Congress: https://www.eurekalert.org/images/2016climateletter6-28-16.pdf]

American Institute of Biological Sciences

[The AIBS is a signatory to the June 28, 2016 letter to the U.S. Congress: https://www.eurekalert.org/images/2016climateletter6-28-16.pdf]

American Institute of Physics 2004

https://web.archive.org/web/20050217173516/http://www.aip.org/fyi/2004/042.html

The Governing Board of the American Institute of Physics has endorsed a position statement on climate change adopted by the American Geophysical Union (AGU) Council in December 2003. AGU is one of ten Member Societies of the American Institute of Physics. The statement follows:

Human Impacts on Climate

Human activities are increasingly altering the Earth's climate. These effects add to natural influences that have been present over Earth's history. Scientific evidence strongly indicates that natural influences cannot explain the rapid increase in global near-surface temperatures observed during the second half of the 20th century.

Human impacts on the climate system include increasing concentrations of atmospheric greenhouse gases (e.g., carbon dioxide, chlorofluorocarbons and their substitutes, methane, nitrous oxide, etc.), air pollution, increasing concentrations of airborne particles, and land alteration. A particular concern is that atmospheric levels of carbon dioxide may be rising faster than at any time in Earth's history, except possibly following rare events like impacts from large extraterrestrial objects…

American Lung Association

The ALA is a signatory to the April 2016 statement: http://www.lung.org/our-initiatives/healthy-air/outdoor/climate-change/declaration-on-climate-change.html?referrer=https://www.google.com/

American Medical Association April 4, 2011

http://www.amednews.com/article/20110404/opinion/304049959/4/

Editorial: Confronting Health Issues of Climate Change

If physicians want evidence of climate change, they may well find it in their own offices. Patients are presenting with illnesses that once happened only in warmer areas. Chronic conditions are becoming aggravated by more frequent and extended heat waves. Allergy and asthma seasons are getting longer. Spates of injuries are resulting from more intense ice storms and snowstorms.

Scientific evidence shows that the world's climate is changing and that the results have public health consequences. The American Medical Association is working to ensure that physicians and others in health care understand the rise in climate-related illnesses and injuries so they can prepare and respond to them. The Association also is promoting environmentally responsible practices that would reduce waste and energy consumption.

April 2016

https://assets.ama-assn.org/sub/advocacy-update/2016-04-28.html

Amicus Brief filed before the Supreme Court in support of the Clean Power Plan.

Failure to uphold the Clean Power Plan would undermine [the] EPA's ability to carry out its legal obligation to regulate carbon emissions that endanger human health and would negatively impact the health of current and future generations.

Carbon emissions are a significant driver of the anthropogenic greenhouse gas emissions that cause climate change and consequently harm human health. Direct impacts from the changing climate include health-related illness, declining air quality and increased respiratory and cardiovascular illness. Changes in climate also facilitate the migration of mosquito-borne diseases, such as dengue fever, malaria and most recently the Zika Virus.

"In surveys conducted by three separate U.S. medical professional societies," the brief said, "a significant majority of surveyed physicians concurred that climate change is occurring … is having a direct impact on the health of their patients, and that physicians anticipate even greater climate-driven adverse human health impacts in the future."

American Meteorological Society August 20, 2012 Statement

[This statement is considered in force until August 2017 unless superseded by a new statement issued by the AMS Council before this date.]

https://www.ametsoc.org/ams/index.cfm/about-ams/ams-statements/statements-of-the-ams-in-force/climate-change/

...Warming of the climate system now is unequivocal, according to many different kinds of evidence. Observations show increases in globally averaged air and ocean temperatures, as well as widespread melting of snow and ice and rising globally averaged sea level. Surface temperature data for Earth as a whole, including readings over both land and ocean, show an increase of about 0.8°C (1.4°F) over the period 1901-2010 and about 0.5°C (0.9°F) over the period 1979–2010 (the era for which satellite-based temperature data are routinely available). Due to natural variability, not every year is warmer than the preceding year globally. Nevertheless, all of the 10 warmest years in the global temperature records up to 2011 have occurred since 1997, with 2005 and 2010 being the warmest two years in more than a century of global records. The warming trend is greatest in northern high latitudes and over land. In the U.S., most of the observed warming has occurred in the West and in Alaska; for the nation as a whole, there have been twice as many record daily high temperatures as record daily low temperatures in the first decade of the 21st century...

There is unequivocal evidence that Earth's lower atmosphere, ocean, and land surface are warming; sea level is rising; and snow cover, mountain glaciers, and Arctic sea ice are shrinking. The dominant cause of the warming since the 1950s is human activities. This scientific finding is based on a large and persuasive body of research. The observed warming will be irreversible for many years into the future, and even larger temperature increases will occur as greenhouse gases continue to accumulate in the atmosphere. Avoiding this future warming will require a large and rapid reduction in global greenhouse gas emissions. The ongoing warming will increase risks and stresses to human societies, economies, ecosystems, and wildlife through the 21st century and beyond, making it imperative that society respond to a changing climate. To inform decisions on adaptation and mitigation, it is critical that we improve our understanding of the global climate system and our ability to project future climate through continued and improved monitoring and research. This is especially true for smaller (seasonal and regional) scales and weather and climate extremes, and for important hydroclimatic variables such as precipitation and water availability…

Technological, economic, and policy choices in the near future will determine the extent of future impacts of climate change. Science-based decisions are seldom made in a context of absolute certainty. National and international policy discussions should include consideration of the best ways to both adapt to and mitigate climate change. Mitigation will reduce the amount of future climate change and the risk of impacts that are potentially large and dangerous. At the same time, some continued climate change is inevitable, and policy responses should include adaptation to climate change. Prudence dictates extreme care in accounting for our relationship with the only planet known to be capable of sustaining human life.

[The AIBS is also a signatory to the June 28, 2016 letter to the U.S. Congress:https://www.eurekalert.org/images/2016climateletter6-28-16.pdf]

American Physical Society November 14, 2015

https://www.aps.org/policy/statements/15_3.cfm

Statement on Earth's Changing Climate

Earth's changing climate is a critical issue and poses the risk of significant environmental, social and economic disruptions around the globe. While natural sources of climate variability are significant, multiple lines of evidence indicate that human influences have had an increasingly dominant effect on global climate warming observed since the mid-twentieth century. Although the magnitudes of future effects are uncertain, human influences on the climate are growing. The potential consequences of climate change are great and the actions taken over the next few decades will determine human influences on the climate for centuries.

On Climate Science:

As summarized in the 2013 report of the Intergovernmental Panel on Climate Change (IPCC), there continues to be significant progress in climate science. In particular, the connection between rising concentrations of atmospheric greenhouse gases and the increased warming of the global climate system is more compelling than ever. Nevertheless, as recognized by Working Group 1 of the IPCC, scientific challenges remain in our abilities to observe, interpret, and project climate changes. To better inform societal choices, the APS urges sustained research in climate science.

On Climate Action:

The APS reiterates its 2007 call to support actions that will reduce the emissions, and ultimately the concentration, of greenhouse gases as well as increase the resilience of society to a changing climate, and to support research on technologies that could reduce the climate impact of human activities. ...

American Psychological Association The APA is a signatory to the April 2016 statement: http://www.lung.org/our-initiatives/healthy-air/outdoor/climate-change/declaration-on-climate-change.html?referrer=https://www.google.com/

American Public Health Association November 03, 2015 Policy Statement 20157

https://www.apha.org/policies-and-advocacy/public-health-policy-statements/policy-database/2015/12/03/15/34/public-health-opportunities-to-address-the-health-effects-of-climate-change

[This policy builds upon and replaces existing policies 20078 (Addressing the Urgent Threat of Global Climate Change to Public Health and the Environment) and 9510 (Global Climate Change)]

Public Health Opportunities to Address the Health Effects of Climate Change

Climate change poses major threats to human health, human and animal populations, ecological stability, and human social, financial, and political stability and well-being. Observed health impacts of climate change include increased heat-related morbidity and mortality, expanded ranges and frequency of infectious disease outbreaks, malnutrition, trauma, violence and political conflict, mental health issues, and loss of community and social connections. Certain populations will experience disproportionate negative effects, including pregnant women, children, the elderly, marginalized groups such as racial and ethnic minorities, outdoor workers, those with chronic diseases, and those in economically disadvantaged communities. Climate change poses significant ethical challenges as well as challenges to global and health equity. The economic risks of inaction may be significant, yet many strategies to combat climate change offer near- and long-term co-benefits to health, producing cost savings that could offset implementation costs. At present, there are major political barriers to adopting strategies to mitigate and adapt to climate change. Recognizing the urgency of the issue and importance of the public health role, APHA, the Centers for Disease Control and Prevention, and others have developed resources and tools to help support public health engagement. APHA calls for individual, community, national, and global action to address the health risks posed by climate change. The public health community has critical roles to play, including advocating for action, especially among policymakers; engaging in health prevention and preparedness efforts; conducting surveillance and research on climate change and health; and educating public health professionals.

[The APHA is also a signatory to the April 2016 statement: http://www.lung.org/our-initiatives/healthy-air/outdoor/climate-change/

declaration-on-climate-change.html?referrer=https://www.google.com/]

[The APHA is also a signatory to the June 28, 2016 letter to the U.S. Congress: https://www.eurekalert.org/images/2016climateletter6-28-16.pdf]

American Quaternary Association September 5, 2006

Letter to EOS of the Council of the AQA

http://onlinelibrary.wiley.com/doi/10.1029/2006EO360008/epdf

The available scientific evidence clearly shows that the Earth on average is becoming warmer... Few credible scientists now doubt that humans have influenced the documented rise of global temperatures since the Industrial Revolution. The first government led U.S. Climate Change Science Program synthesis and assessment report supports the growing body of evidence that warming of the atmosphere, especially over the past 50 years, is directly impacted by human activity.

American Society for Microbiology 2003 Global Environmental Change Statement

http://www.asm.org/images/docfilename/0000006005/globalwarming%5B1%5D.pdf

In 2003, the ASM issued a policy report in which they recommend "reducing net anthropogenic CO_2 emissions to the atmosphere" and "minimizing anthropogenic disturbances of" atmospheric gases:

"Carbon dioxide concentrations were relatively stable for the past 10,000 years but then began to increase rapidly about 150 years ago... as a result of fossil fuel consumption and land use change. Of course, changes in atmospheric composition are but one component of global change, which also includes disturbances in the physical and chemical conditions of the oceans and land surface. Although global change has been a natural process throughout Earth's history, humans are responsible for substantially accelerating present-day changes. These changes may adversely affect human health and the biosphere on which we depend. Outbreaks of a number of diseases, including Lyme disease, hantavirus infections, dengue fever, bubonic plague, and cholera, have been linked to climate change."

American Society of Agronomy May 2011

https://www.soils.org/files/science-policy/asa-cssa-sssa-climate-change-policy-statement.pdf

A comprehensive body of scientific evidence indicates beyond reasonable doubt that global climate change is now occurring and that its manifestations threaten the stability of societies as well as natural and managed ecosystems. Increases in ambient temperatures and changes in related processes are directly linked to rising anthropogenic greenhouse gas (GHG) concentrations in the atmosphere. The potential related impacts of climate change on the ability of agricultural systems, which include soil and water resources, to provide food, feed, fiber, and fuel, and maintenance of ecosystem services (e.g., water supply and habitat for crop landraces, wild relatives, and pollinators) as well as the integrity of the environment, are major concerns.

Around the world and in the United States (US), agriculture—which is comprised of field, vegetable, and tree crops, as well as livestock production—constitutes a major land use which influences global ecosystems. Globally, crop production occupies approximately 1.8 Billion (B) hectares out of a total terrestrial land surface of about 13.5 B hectares. In addition, animal production utilizes grasslands, rangelands, and savannas, which altogether cover about a quarter of the Earth's land. Even in 2010, agriculture remains the most basic and common human occupation on the planet and a major contributor to human well-being.

Changes in climate are already affecting the sustainability of agricultural systems and disrupting production.

[The May 2011 statement was also signed by the Crop Science Society of America and the Soil Science Society of America.]

[The ASoA is also a signatory to the June 28, 2016 letter to the U.S. Congress: https://www.eurekalert.org/images/2016climateletter6-28-16.pdf]

American Society of Civil Engineers July 18, 2015 Policy Statement 360

There is strong evidence that the climate is changing and will continue to change. Climate scientists project that there will be substantial increases in temperature with related increases in atmospheric water vapor and increases in extreme precipitation amounts and intensities in most geographic regions as a result of climate change. However, while there is clear evidence of a changing climate, understanding the significance of climate change at the temporal and spatial scales as it relates to engineering practice is more difficult.

There is an increasing demand for engineers to address future climate change into project design criteria; however, current practices and rules governing such practices do not adequately address concerns associated with climate change…

Climate change poses a potentially serious impact on worldwide water resources, energy production and use, agriculture, forestry, coastal development and resources, flood control and public infrastructure…

The American Society of Civil Engineers (ASCE) supports:

- Government policies that encourage anticipation of and preparation for impacts of climate change on the built environment.

- Revisions to engineering design standards, codes, regulations and associated laws that govern infrastructure potentially affected by climate change.
- Research, development and demonstration to advance recommended civil engineering practices and standards to effectively address climate change impacts.
- Cooperative research involving engineers with climate, weather, and life scientists to gain a better understanding of the magnitudes and consequences of future extremes.
- Informing practicing engineers, project stakeholders, policy makers and decision makers about the uncertainty in projecting future climate and the reasons for the uncertainty.
- Developing a new paradigm for engineering practice in a world in which climate is changing but the extent and time of local impacts cannot be projected with a high degree of certainty.
- Identifying critical infrastructure that is most threatened by changing climate in a given region and informing decision makers and the public.

American Society of Ichthyologists and Herpetologists

The ASIH is a signatory to the June 28, 2016 letter to the U.S. Congress: https://www.eurekalert.org/images/2016climateletter6-28-16.pdf

American Society of Naturalists The ASN is a signatory to the June 28, 2016 letter to the U.S. Congress: https://www.eurekalert.org/images/2016climateletter6-28-16.pdf

American Society of Plant Biologists

[The ASPB is a signatory to the June 28, 2016 letter to the U.S. Congress: https://www.eurekalert.org/images/2016climateletter6-28-16.pdf]

American Statistical Association November 30, 2007

Adopted by the ASA Board of Directors

https://web.archive.org/web/20130307002012/http://www.amstat.org/news/climatechange.cfm

ASA Statement on Climate Change

The American Statistical Association (ASA) recently convened a workshop of leading atmospheric scientists and statisticians involved in climate change research. The goal of this workshop was to identify a consensus on the role of statistical science in current assessments of global warming and its impacts. Of particular interest to this workshop was the recently published Fourth Assessment Report of the United Nations' Intergovernmental Panel on Climate Change (IPCC), endorsed by more than 100 governments and drawing on the expertise of a large portion of the climate science community.

Through a series of meetings spanning several years, IPCC drew in leading experts and assessed the relevant literature in the geosciences and related disciplines as it relates to climate change. The Fourth Assessment Report finds that "Warming of the climate system is unequivocal, as is now evident from observations of increases in global average air and ocean temperatures, widespread melting of snow and ice, and rising mean sea level. … Most of the observed increase in globally averaged temperatures since the mid-20th century is very likely due to the observed increase in anthropogenic greenhouse gas concentrations. … Discernible human influences now extend to other aspects of climate, including ocean warming, continental-average temperatures, temperature extremes, and wind patterns.

The ASA endorses the IPCC conclusions.

[The ASA is also a signatory to the June 28, 2016 letter to the U.S. Congress: https://www.eurekalert.org/images/2016climateletter6-28-16.pdf]

American Water Resources Association August 5, 2015

http://www.awra.org/policy/policy-statements-leveraging-IWRM.html

After people, water is our most critical and strategic natural resource, yet the U.S. lack a national strategy for water resources management. In addition, Americans are the world's largest water consumers. Threats of an aging infrastructure, climate change and population growth are so significant that the nation can no longer afford to postpone action. It's imperative that a focused effort be articulated and initiated to create and demonstrate strategies to sustain U.S. water resources. The country's future growth and prosperity depend on it.

American Thoracic Society

The ATS is also a signatory to the April 2016 statement: http://www.lung.org/our-initiatives/healthy-air/outdoor/climate-change/declaration-on-climate-change.html?referrer=https://www.google.com/

Association for the Sciences of Limnology and Oceanography

The ASLO is a signatory to the June 28, 2016 letter to the U.S. Congress: https://www.eurekalert.org/images/2016climateletter6-28-16.pdf

Association for Tropical Biology and Conservation

The ATBC is a signatory to the June 28, 2016 letter to the U.S. Congress: https://www.eurekalert.org/images/2016climateletter6-28-16.pdf

Association of Ecosystem Research Centers

The AERC is a signatory to the June 28, 2016 letter to the U.S. Congress: https://www.eurekalert.org/images/2016climateletter6-28-16.pdf

Asthma and Allergy Foundation of America

The AAFA is a signatory to the April 2016 statement: http://www.lung.org/our-initiatives/healthy-air/outdoor/climate-change/declaration-on-climate-change.html?referrer=https://www.google.com/

Australian Coral Reef Society June 16, 2006

https://web.archive.org/web/20060322170802/http://www.australiancoralreefsociety.org/pdf/chadwick605a.pdf

There is broad scientific consensus that coral reefs are heavily affected by the activities of man and there are significant global influences that can make reefs more vulnerable such as global warming... It is highly likely that coral bleaching has been exacerbated by global warming.

There is almost total consensus among experts that the earth's climate is changing as a result of the build-up of greenhouse gases. The IPCC (involving over 3,000 of the world's experts) has come out with clear conclusions as to the reality of this phenomenon. One does not have to look further than the collective academy of scientists worldwide to see the string (of) statements on this worrying change to the earth's atmosphere…

September 1, 2016

http://www.australiancoralreefsociety.org/c/document_library/get_file?uuid=2e52d369-95a0-451a-863d-5cae22ed625e&groupId=10136

Science-based policy plan for the Great Barrier Reef

Discussion: Advancing Climate Action in Queensland

Given the observed damage caused by a temperature increase of ~1°C above pre-industrial levels, we urge all possible actions to keep future warming below the 1.5°C target set by the Paris Agreement. The following proposed initiatives will act to reduce the severity of climate-inflicted damage on reefs, helping to avoid total ecological collapse. The ACRS strongly supports the following proposed actions…

Australian Institute of Physics March 10, 2005, Policy Document 1.01

https://web.archive.org/web/20080201000000*/http://www.aip.org.au/scipolicy/Science%20Policy.pdf

The AIP supports a reduction of the green house gas emissions that are leading to increased global temperatures, and encourages research that works towards this goal…

Research in Australia and overseas shows that an increase in global temperature will adversely affect the Earth's climate patterns. The melting of the polar ice caps, combined with thermal expansion, will lead to rises in sea levels that may impact adversely on our coastal cities. The impact of these changes on biodiversity will fundamentally change the ecology of Earth…

Australian Medical Association August 28, 2015

https://ama.com.au/position-statement/ama-position-statement-climate-change-and-human-health-2004-revised-2015

Human health is ultimately dependent on the health of the planet and its ecosystem. The AMA recognises the latest findings regarding the science of climate change, the role of humans, past observations and future projections. The consequences of climate change have serious direct and indirect, observed and projected health impacts both globally and in Australia. There is inequity in the distribution of these health impacts both within and between countries, with some groups being particularly vulnerable. In recognition of these issues surrounding climate change and health, the AMA believes that:

- because climate change involves potentially serious or irreversible harm to the environment and to human health, urgent international cooperation is essential to mitigate climate change. Reducing greenhouse gas emissions within a global carbon budget is necessary to prevent further climate harm as a result of human activity.

- Australia should adopt mitigation targets within an Australian carbon budget that represents Australia's fair share of global greenhouse gas emissions, under the principle of common but differential responsibilities.
- climate policies can have public health benefits beyond their intended impact on the climate. These health benefits should be promoted as a public health opportunity, with significant potential to offset some costs associated with addressing climate change.
- the health impacts of climate change and the health co-benefits of climate mitigation policies both bear economic costs and savings. Economic evaluations of the costs and benefits of climate policies must therefore incorporate the predicted public health impact accrued from such policies and the public health costs of unmitigated climate change.
- Regional and national collaboration across all sectors, including a comprehensive and broad reaching adaptation plan is necessary to reduce the health impacts of climate change. This requires a National Strategy for Health and Climate Change.
- there should be greater education and awareness of the health impacts of climate change, and the public health benefits of mitigation and adaptation.
- renewable energy presents relative benefits compared to fossil fuels with regard to air pollution and health. Therefore, active transition from fossil fuels to renewable energy sources should be considered.
- decarbonisation of the economy can potentially result in unemployment and subsequent adverse health impacts. The tran-

sition of workers displaced from carbon intensive industries must be effectively managed.

Australian Meteorological and Oceanographic Society February 2, 2016

http://www.amos.org.au/Main/About_us/Statements/Main/Statements.aspx?hkey=2f71e26e-372c-41b2-a1e0-700d89c3d4f5

Statement on Climate Change

Global climate has changed substantially. Global climate change and global warming are real and observable…

Human influence has been detected in the warming of the atmosphere and the ocean globally, and in Australia. It is now certain that the human activities that have increased the concentration of greenhouse gases in the atmosphere contribute significantly to observed warming. Further it is extremely likely that these human activities are responsible for most of the observed global warming since 1950. The warming associated with increases in greenhouse gases originating from human activity is called the enhanced greenhouse effect….

Our climate is very likely to continue to change as a result of human activity.

Global temperature increases are already set to continue until at least the middle of this century even if emissions were reduced to zero. The magnitude of warming and related changes can be limited depending on the total amount of carbon dioxide and other greenhouse gases ultimately emitted as a result of human activities; future climate scenarios depend critically on future changes in emissions…

BioQUEST Curriculum Consortium

BioQUEST is a signatory to the June 28, 2016 letter to the U.S. Congress: https://www.eurekalert.org/images/2016climateletter6-28-16.pdf

Botanical Society of America

The BSA is a signatory to the June 28, 2016 letter to the U.S. Congress: https://www.eurekalert.org/images/2016climateletter6-28-16.pdf

Canadian Foundation for Climate and Atmospheric Sciences

November 25, 2005

https://scentofpine.files.wordpress.com/2011/05/cfcas-letter-to-canadian-prime-minister-paul-martin-nov-2005.pdf

We, the members of the Board of Trustees of CFCAS and Canadian climate science leaders from the public and academic sectors in Canada, concur with The Joint Science Academies statement that *"climate change is real"* and note that the 2004 Arctic Climate Impact Assessment concluded that Arctic temperatures have risen at almost twice the rate of the rest of the world over the past few decades. Furthermore, we endorse the assessment of climate science undertaken by the Intergovernmental Panel on Climate Change (IPCC) and its conclusion that "There is new and stronger evidence that most of the warming observed over the last 50 years is attributable to human activities."

There is now increasing unambiguous evidence of a changing climate in Canada and around the world... There is an increasing urgency to act on the threat of climate change. Significant steps are needed to stop the growth in atmospheric greenhouse gas concentrations by reducing emissions. Since mitigation measures will become effective only after many years, adaptive strategies as well are of great importance and need to begin now....

Canadian Meteorological and Oceanographic Society 2013-2014

http://www.cmos.ca/site/ps_pos_statements?a=7

Updated Statement on Human-Induced Climate Change

...Since the industrial revolution of the early 19th century, human activities have also markedly influenced the climate. This well-documented human-induced change is large and very rapid in comparison to past changes in the Earth's climate...

Even if the human-induced emission of greenhouse gases into the atmosphere were to cease today, past emissions have committed the world to long-term changes in climate. Carbon dioxide emitted from the combustion of fossil fuels will remain in the atmosphere for centuries to millennia, and the slow ocean response to atmospheric warming will cause the climate change to persist even longer. Further CO2 emissions will lead to greater human-induced change in proportion to total cumulative emissions. Meaningful interventions to mitigate climate change require a reduction in emissions. To avoid societally, economically, and ecologically disruptive changes to the Earth's climate, we will have little choice but to leave much of the unextracted fossil fuel carbon in the ground...

The urgent challenges for the global community, and Canadians in particular, are to learn how to adapt to the climate changes to which we are already committed and to develop effective and just responses to avoid further damaging climate change impacts for both present and future generations.

Consortium for Ocean Leadership

The COL is a signatory to the June 28, 2016 letter to the U.S. Congress: https://www.eurekalert.org/images/2016climateletter6-28-16.pdf

Crop Science Society of America May 2011

https://www.soils.org/files/science-policy/asa-cssa-sssa-climate-change-policy-statement.pdf

A comprehensive body of scientific evidence indicates beyond reasonable doubt that global climate change is now occurring and that its manifestations threaten the stability of societies as well as natural and managed ecosystems. Increases in ambient temperatures and changes in related processes are directly linked to rising anthropogenic greenhouse gas (GHG) concentrations in the atmosphere. The potential related impacts of climate change on the ability of agricultural systems, which include soil and water resources, to provide food, feed, fiber, and fuel, and maintenance of ecosystem services (e.g., water supply and habitat for crop landraces, wild relatives, and pollinators) as well as the integrity of the environment, are major concerns.

Around the world and in the United States (US), agriculture—which is comprised of field, vegetable, and tree crops, as well as livestock production—constitutes a major land use which influences global ecosystems. Globally, crop production occupies approximately 1.8 Billion (B) hectares out of a total terrestrial land surface of about 13.5 B hectares. In addition, animal production utilizes grasslands, rangelands, and savannas, which altogether cover about a quarter of the Earth's land. Even in 2010, agriculture remains the most basic and common human occupation on the planet and a major contributor to human well-being. Changes in climate are already affecting the sustainability of agricultural systems and disrupting production.

[The May 2011 Statement was also signed by the American Society of Agronomy and the Soil Science Society of America.]

[The CSSA is also a signatory to the June 28, 2016 letter to the U.S. Congress: https://www.eurekalert.org/images/2016climateletter6-28-16.pdf]

Ecological Society of America (As of January 2017)

http://www.esa.org/esa/esa-position-statement-ecosystem-management-in-a-changing-climate/

ESA Position Statement: Ecosystem Management in a Changing Climate

Ecosystems are already responding to climate change. Continued warming—some of which is now unavoidable—may impair the ability of many such systems to provide critical resources and services like food, clean water, and carbon sequestration. Buffering against the impacts of climate change will require new strategies to both mitigate the extent of change and adapt to changes that are inevitable. The sooner such strategies are deployed, the more effective they will be in reducing irreversible damage.

Ecosystems can be managed to limit and adapt to both the near- and long-term impacts of climate change. Strategies that focus on restoring and maintaining natural ecosystem function (reducing deforestation, for example) are the most prudent; strategies that drastically alter ecosystems may have significant and unpredictable impacts…

The Reality of Climate Change

The Earth is warming— average global temperatures have increased by 0.74°C (1.3°F) in the past 100 years. The scientific community agrees that catastrophic and possibly irreversible environmental change will occur if average global temperatures rise an additional 2°C (3.6°F). Warming to date has already had significant impacts on the Earth and its ecosystems, including increased droughts, rising sea levels, disappearing glaciers, and changes in the distribution and seasonal activities of many species…

The Source of Climate Change

Most warming seen since the mid 1900s is very likely due to greenhouse gas emissions from human activities. Global emissions have risen rapidly since pre-industrial times, increasing 70% between 1970 and 2004 alone…

The Future of Climate Change:

Even if greenhouse gas emissions stop immediately, global temperatures will continue to rise at least for the next 100 years. Depending on the extent and effectiveness of climate change mitigation strategies, global temperatures could rise 1-6°C (2-10°F) by the end of the 21st century, according to the Intergovernmental Panel on Climate Change. Swift and significant emissions reductions will be vital in minimizing the impacts of warming…

[The ESA is also a signatory to the June 28, 2016 letter to the U.S. Congress: https://www.eurekalert.org/images/2016climateletter6-28-16.pdf]

Engineers Australia (The Institution of Engineers Australia)
November 2014

https://www.engineersaustralia.org.au/sites/default/files/climate_change_policy_nov_2014.pdf

Engineers Australia accepts the comprehensive scientific basis regarding climate change, the influence of anthropogenic global warming, and that climate change can have very serious community consequences.

Engineers are uniquely placed to provide both mitigation and adaptation solutions for this serious global problem, as well as address future advances in climate change science.

This Climate Change Policy Statement has been developed to enable organisational governance on the problem, and provide support for members in the discipline and practice of the engineering profession.

Context

Building upon a long history of Engineers Australia policy development, and as the largest technically informed professional body in Australia, Engineers Australia advocates that Engineers must act proactively to address climate change as an ecological, social and economic risk...

Entomological Society of America

The ESA is also a signatory to the June 28, 2016 letter to the U.S. Congress: https://www.eurekalert.org/images/2016climateletter6-28-16.pdf

European Academy of Sciences and Arts

March 3, 2007

http://www.euro-acad.eu/downloads/memorandas/lets_be_honest_-_festplenum_03.03.07_-_final2.pdf

Human activity is most likely responsible for climate warming. Most of the climatic warming over the last 50 years is likely to have been caused by increased concentrations of greenhouse gases in the atmosphere. Documented long-term climate changes include changes in Arctic temperatures and ice, widespread changes in precipitation amounts, ocean salinity, wind patterns and extreme weather including droughts, heavy precipitation, heat waves and the intensity of tropical cyclones. The above development potentially has dramatic consequences for mankind's future...

European Federation of Geologists January 23, 2008

http://eurogeologists.eu/wp-content/uploads/2015/10/Position-Paper_Carbon-Capture-and-geological-Storage.pdf

The EFG recognizes the work of the IPCC and other organizations, and subscribes to the major findings that climate change is happening, is predominantly caused by anthropogenic emissions of CO2, and poses a significant threat to human civilization. Anthropogenic CO2 emissions come from fossil carbon sources, such as coal, oil, natural gas, limestone and carbonate rocks. Thriving and developing economies currently depend on these resources. Since geologists play a crucial role in their exploration and exploitation, we feel praised by the increasing welfare, but also implicated by the carbon curse.

It is clear that major efforts are necessary to quickly and strongly reduce CO2 emissions. The EFG strongly advocates renewable and sustainable energy production, including geothermal energy, as well as the need for increasing energy efficiency.

European Geosciences Union November 2008

http://www.egu.eu/about/statements/egu-position-statement-on-ocean-acidification/

EGU position statement on climate/ocean acidification

Impacts of ocean acidification may be just as dramatic as those of global warming (resulting from anthropogenic activities on top of natural variability) and the combination of both are likely to exacerbate consequences, resulting in potentially profound changes throughout marine ecosystems and in the services that they provide to humankind...

Since the beginning of the industrial revolution the release of carbon dioxide (CO_2) from our industrial and agricultural activities has resulted in atmospheric CO_2 concentrations that have increased from approximately 280 to 385 parts per million (ppm). The atmospheric concentration of CO_2 is now higher than experienced on Earth for at least the last 800,000 years (direct ice core evidence) and probably the last 25 million years, and is expected to continue to rise at an increasing rate, leading to significant temperature increases in the atmosphere and ocean in the coming decades…

Ocean acidification is already occurring today and will continue to intensify, closely tracking atmospheric CO2 increase. Given the potential threat to marine ecosystems and its ensuing impact on human society and economy, especially as it acts in conjunction with anthropogenic global warming, there is an urgent need for immediate action.

This rather new recognition that, in addition to the impact of CO_2 as a greenhouse gas on global climate change, OA is a **direct** consequence of the absorption of anthropogenic CO_2 emissions, will hopefully help to set in motion an even more stringent CO_2 mitigation policy worldwide. The only solutions to avoid excessive OA are a long-term mitigation strategy to limit future release of CO_2 to the atmosphere and/or enhance removal of excess CO_2 from the atmosphere.

European Physical Society November 2007

http://archive.iupap.org/epspositionpaper.pdf

The emission of anthropogenic greenhouse gases, among which carbon dioxide is the main contributor, has amplified the natural greenhouse effect and led to global warming. The main contribution stems from burning fossil fuels. A further increase will have decisive effects on life on earth. An energy cycle with the lowest possible CO2 emission is called for wherever possible to combat climate change.

2015 Statement

http://www.eps.org/resource/resmgr/policy/eps-pp-EuropeanEnergyPol2015.pdf

The forthcoming United Nations Climate Change Conference (Paris, December 2015) will be held with the objective of achieving a binding and global agreement on climate-related policy from all nations of the world. This conference, seeking to protect the climate, will be a great opportunity to find solutions in the human quest for sustainable energy as a global endeavour. The Energy Group of the European Physical Society (EPS) welcomes the energy policy of the European Union (EU) to promote renewable energies for electricity generation, together with energy efficiency measures. This policy needs to be implemented by taking into account the necessary investments and the impact on the economical position of the EU in the world. Since the direct impact of any EU energy policy on world CO2 emissions is rather limited, the best strategy is to take the lead in mitigating climate change and in developing an energy policy that offers an attractive and economically viable model with reduced CO2 emissions and lower energy dependence...

European Science Foundation 2007

http://archives.esf.org/fileadmin/Public_documents/Publications/MB_Climate_Change_Web.pdf

The scientific evidence is now overwhelming that climate change is a serious global threat which requires an urgent global response, and that climate change is driven by human activity... Enough is now known to make climate change the challenge of the 21st century, and the research community is poised to address this challenge...

There is now convincing evidence that since the industrial revolution, human activities, resulting in increasing concentrations of greenhouse gases have become a major agent of climate change. These greenhouse gases affect the global climate by retaining heat in the troposphere, thus raising the average temperature of the planet and altering global atmospheric circulation and precipitation patterns.

While on-going national and international actions to curtail and reduce greenhouse gas emissions are essential, the levels of greenhouse gases currently in the atmosphere, and their impact, are likely to persist for several decades. On-going and increased efforts to mitigate climate change through reduction in greenhouse gases are therefore crucial...

European Space Sciences Committee December 2015

http://archives.esf.org/media-centre/ext-single-news/article/essc-statement-on-climate-change-1096.html

The European Space Sciences Committee (ESSC) supports the Article (2) agreement on climate change of the Declaration of the '2015 Budapest World Science Forum on the enabling power of science' urges such a universal agreement aiming at stabilising atmospheric concentrations of greenhouse gases and reducing the amount of airborne particles.

The ESSC encourages countries to reduce their emissions in order to avoid dangerous anthropogenic interference with the climate system, which could lead to disastrous consequences. Such consequences, albeit from natural evolution, are witnessed in other objects of our Solar System.

Federation of Australian Scientific and Technological Societies
September 4, 2008

https://scentofpine.files.wordpress.com/2012/09/fasts-statement-on-climate-change-sep-2008.pdf

Global climate change is real and measurable. Since the start of the 20th century, the global mean surface temperature of the Earth has increased by more than 0.7°C and the rate of warming has been largest in the last 30 years… Key vulnerabilities arising from climate change include water resources, food supply, health, coastal settlements, biodiversity and some key ecosystems such as coral reefs and alpine regions. As the atmospheric concentration of greenhouse gases increases, impacts become more severe and widespread. To reduce the global net economic, environmental and social losses in the face of these impacts, the policy objective must remain squarely focused on returning greenhouse gas concentrations to near pre-industrial levels through the reduction of emissions… The spatial and temporal fingerprint of warming can be traced to increasing greenhouse gas concentrations in the atmosphere, which are a direct result of burning fossil fuels, broad-scale deforestation and other human activity.

Geological Society of America

Adopted in October 2006; revised April 2010; March 2013; April 2015

Decades of scientific research have shown that climate can change from both natural and anthropogenic causes. The Geological Society of America (GSA) concurs with assessments by the National Academies of Science (2005), the National Research Council (2011), the Intergovernmental Panel on Climate Change (IPCC, 2013) and the U.S. Global Change Research Program (Melillo et al., 2014) that global climate has warmed in response to increasing concentrations of carbon dioxide (CO2) and other greenhouse gases. The concentrations of greenhouse gases in the atmosphere are now higher than they have been for many thousands of years. Human activities (mainly greenhouse-gas emissions) are the dominant cause of the rapid warming since the middle 1900s (IPCC, 2013). If the upward trend in greenhouse-gas concentrations continues, the projected global climate change by the end of the twenty-first century will result in significant impacts on humans and other species. The tangible effects of climate change are already occurring. Addressing the challenges posed by climate change will require a combination of adaptation to the changes that are likely to occur and global reductions of CO2 emissions from anthropogenic sources…

[The GSA is also a signatory to the June 28, 2016 letter to the U.S. Congress: https://www.eurekalert.org/images/2016climateletter6-28-16.pdf]

Health Care Without Harm

The HCWH is a signatory to the April 2016 statement: http://www.lung.org/our-initiatives/healthy-air/outdoor/climate-change/declaration-on-climate-change.html?referrer=https://www.google.com/

Health Care Climate Council

The HCCC is a signatory to the April 2016 statement: http://www.lung.org/our-initiatives/healthy-air/outdoor/climate-change/declaration-on-climate-change.html?referrer=https://www.google.com/

Institute of Professional Engineers (New Zealand) 2001

https://web.archive.org/web/20080815000000*/http://www.ipenz.org.nz/ipenz/forms/pdfs/Info_Note_6.pdf

Human activities have increased the concentration of these atmospheric greenhouse gases, and although the changes are relatively small, the equilibrium maintained by the atmosphere is delicate, and so the effect of these changes is significant. The world's most important greenhouse gas is carbon dioxide, a by-product of the burning of fossil fuels.

... Professional engineers commonly deal with risk, and frequently have to make judgments based on incomplete data. The available evidence suggests very strongly that human activities have already begun to make significant changes to the earth's climate, and that the longterm risk of delaying action is greater than the cost of avoiding/minimising the risk.

Inter-Academy Council 2007

http://www.interacademycouncil.net/24026/25142.aspx

Scientific evidence is overwhelming that current energy trends are unsustainable.

Immediate action is required to effect change in the timeframe needed to address significant ecological, human health and development, and energy security needs. Aggressive changes in policy are thus needed to accelerate the deployment of superior technologies. With a combination of such policies at the local, national, and international level, it should be possible—both technically and economically—to elevate the living conditions of most of humanity, while simultaneously addressing the risks posed by climate change and other forms of energy-related environmental degradation and reducing the geopolitical tensions and economic vulnerabilities generated by existing patterns of dependence on predominantly fossil-fuel resources…

The Study Panel believes that, given the dire prospect of climate change, the following three recommendations should be acted upon *without delay and simultaneously*:

- Concerted efforts should be mounted to improve energy efficiency and reduce the carbon intensity of the world economy, including the worldwide introduction of price signals for carbon emissions, with consideration of different economic and energy systems in individual countries.
- Technologies should be developed and deployed for capturing and sequestering carbon from fossil fuels, particularly coal.
- Development and deployment of renewable energy technologies should be accelerated in an environmentally responsible way.

Taking into account the three urgent recommendations above, another recommendation stands out by itself as a moral and social imperative and should be pursued with all means available

International Association for Great Lakes Research
February 25, 2009

http://iaglr.org/scipolicy/factsheets/iaglr_crossroads_climatechange.pdf

While the Earth's climate has changed many times during the planet's history because of natural factors, including volcanic eruptions and changes in the Earth's orbit, never before have we observed the present rapid rise in temperature and carbon dioxide (CO_2).

Human activities resulting from the industrial revolution have changed the chemical composition of the atmosphere....

Deforestation is now the second largest contributor to global warming, after the burning of fossil fuels. These human activities have significantly increased the concentration of "greenhouse gases" in the atmosphere...

As the Earth's climate warms, we are seeing many changes: stronger, more destructive hurricanes; heavier rainfall; more disastrous flooding; more areas of the world experiencing severe drought; and more heat waves.

International Council of Academies of Engineering and Technological Sciences 2007

http://www.caets.org/cms/7122/7735.aspx

As reported by the Intergovernmental Panel on Climate Change (IPCC), most of the observed global warming since the mid-20th century is very likely due to human-produced emission of greenhouse gases and this warming will continue unabated if present anthropogenic emissions continue or, worse, expand without control. CAETS, therefore, endorses the many recent calls to decrease and control greenhouse gas emissions to an acceptable level as quickly as possible.

International Union for Quaternary Research

http://www.inqua.org/files/iscc.pdf

Climate change is real

There is now strong evidence that significant global warming is occurring. The evidence comes from direct measurements of rising surface air temperatures and subsurface ocean temperatures and, indirectly, from increases in average global sea levels, retreating glaciers, and changes in many physical and biological systems. It is very likely that most of the observed increase in global temperatures since the mid-twentieth century is due to human-induced increases in greenhouse gas concentrations in the atmosphere (IPCC 2007). Human activities are now causing atmospheric concentrations of greenhouse gases – including carbon dioxide, methane, tropospheric ozone, and nitrous oxide – to rise well above pre-industrial levels.

Carbon dioxide levels have increased from 280 ppm in 1750 to over 380 ppm today, higher than any previous levels in at least the past 650,000 years. Increases in greenhouse gases are causing temperatures to rise; the Earth's surface warmed by approximately 0.6°C over the twentieth century. The Intergovernmental Panel on Climate Change (IPCC) has forecast that average global surface temperatures will continue to increase, reaching between 1.1°C and 6.4°C above 1990 levels, by 2100.

The uncertainties about the amount of global warming we face in coming decades can be reduced through further scientific research. Part of this research must be better documenting and understanding past climate change. Research on Earth's climate in the recent geologic past provides insights into ways in which climate can change in the future. It also provides data that contribute to the testing and improvement of the computer models that are used to predict future climate change.

Reduce the causes of climate change

The scientific understanding of climate change is now sufficiently clear to justify nations taking prompt action. A lack of full scientific certainty about some aspects of climate change is not a reason for delaying an immediate response that will, at a reasonable cost, prevent dangerous anthropogenic interference with the climate system. It is vital that all nations identify cost-effective steps that they can take now to contribute to substantial and long-term reduction in net global greenhouse gas emissions. Action taken now to reduce significantly the build-up of greenhouse gases in the atmosphere will lessen the magnitude and rate of climate change. Fossil fuels, which are responsible for most of carbon dioxide emissions produced by human activities, provide valuable resources for many nations and will provide 85% of the world energy demand over the next 25 years (IEA 2004). Minimizing the amount of this carbon dioxide reaching the atmosphere presents a huge challenge but must be a global priority.

International Union of Geodesy and Geophysics July 2007

Resolution 6: The Urgency of Addressing Climate Change

Considering,

The advances in scientific understanding of the Earth system generated by collaborative international, regional, and national observations and research programs; and

The comprehensive and widely accepted and endorsed scientific assessments carried out by the Intergovernmental Panel on Climate Change and regional and national bodies, which have firmly established, on the basis of scientific evidence, that human activities are the primary cause of recent climate change;

Realizing,

Continuing reliance on combustion of fossil fuels as the world's primary source of energy will lead to much higher atmospheric concentrations of greenhouse gases, which will, in turn, cause significant increases in surface temperature, sea level, ocean acidification, and their related consequences to the environment and society;

Stabilization of climate to avoid "dangerous anthropogenic interference with the climate system", as called for in the UN Framework Convention on Climate Change, will require significant cutbacks in greenhouse gas emissions during the 21st century; and

Mitigation of and adaptation to climate change can be made more effective by reducing uncertainties regarding feedbacks and the associated mechanisms;

Urges,

Nations collectively to begin to reduce sharply global atmospheric emissions of greenhouse gases and absorbing aerosols, with the goal of urgently halting their accumulation in the atmosphere and holding atmospheric levels at their lowest practicable value;

National and international agencies to adequately support comprehensive observation and research programs that can clarify the urgency and extent of needed mitigation and promote adaptation to the consequences of climate change;

Resource managers, planners, and leaders of public and private organizations to incorporate information on ongoing and projected changes in climate and its ramifications into their decision-making, with goals of limiting emissions, reducing the negative consequences of climate change, and enhancing adaptation, public well-being, safety, and economic vitality; and

Organizations around the world to join with IUGG and its member Associations to encourage scientists to communicate freely and widely with public and private decision-makers about the consequences and risks of on-going climate change and actions that can be taken to limit climate change and promote adaptation; and

Resolves,

To act with its member Associations to develop and implement an integrated communication and outreach plan to increase public understanding of the nature and implications of human-induced impacts on the Earth system, with the aim of reducing detrimental consequences.

London Mathematical Society

The LMS is a signatory to the July 21, 2015 UK science communiqué on climate change

https://royalsociety.org/~/media/policy/Publications/2015/21-07-15-climate-communique.PDF

National Association of County and City Health Officials

The NACCHO is a signatory to the April 2016 declaration: http://www.lung.org/our-initiatives/healthy-air/outdoor/climate-change/declaration-on-climate-change.html?referrer=https://www.google.com/

National Association of Geoscience Teachers November 10, 2008

http://nagt.org/nagt/policy/ps-climate.html

The National Association of Geoscience Teachers (NAGT) recognizes: (1) that Earth's climate is changing, (2) that present warming trends are largely the result of human activities, and (3) that teaching climate change science is a fundamental and integral part of earth science education. The core mission of NAGT is to "*foster improvement in the teaching of the earth sciences at all levels of formal and informal instruction, to emphasize the cultural significance of the earth sciences and to disseminate knowledge in this field to the general public.*" The National Science Education Standards call for a populace that understands how scientific knowledge is both generated and verified, and how complex interactions between human activities and the environment can impact the Earth system. Climate is clearly an integral part of the Earth system connecting the physical, chemical and biological components and playing an essential role in how the Earth's environment interacts with human culture and societal development. Thus, climate change science is an essential part of Earth Science education and is fundamental to the mission set forth by NAGT. In recognition of these imperatives, NAGT strongly supports and will work to promote education in the science of climate change, the causes and effects of current global warming, and the immediate need for policies and actions that reduce the emission of greenhouse gases.

National Association of Hispanic Nurses

The NAHN is a signatory to the April 2016 declaration: http://www.lung.org/our-initiatives/healthy-air/outdoor/climate-change/declaration-on-climate-change.html?referrer=https://www.google.com/

National Association of Marine Laboratories

The NAML is a signatory to the June 28, 2016 letter to the U.S. Congress: https://www.eurekalert.org/images/2016climateletter6-28-16.pdf

National Environmental Health Association

The NEHA is a signatory to the April 2016 declaration: http://www.lung.org/our-initiatives/healthy-air/outdoor/climate-change/declaration-on-climate-change.html?referrer=https://www.google.com/

National Medical Association

The NMA is a signatory to the April 2016 declaration: http://www.lung.org/our-initiatives/healthy-air/outdoor/climate-change/declaration-on-climate-change.html?referrer=https://www.google.com/

National Academies of Science (selected joint statements)

Many national science academies have published formal statements and declarations acknowledging the state of climate science, the fact that climate is changing, the compelling evidence that humans are responsible, and the need to debate and implement strategies to reduce emissions of greenhouse gases. A few examples of joint academy statements are listed here.

2001

http://science.sciencemag.org/content/292/5520/1261

Following the release of the third in the ongoing series of international reviews of climate science conducted by the Intergovernmental Panel on Climate Chang (IPCC), seventeen national science academies issued a joint statement, entitled "The Science of Climate Change," acknowledging the IPCC study to be the scientific consensus on climate change science.

The statement was signed by: Australian Academy of Sciences, Royal Flemish Academy of Belgium for Sciences and the Arts, Brazilian Academy of Sciences, Royal Society of Canada, Caribbean Academy of Sciences, Chinese Academy of Sciences, French Academy of Sciences, German Academy of Natural Scientists Leopoldina, Indian National Science Academy, Indonesian Academy of Sciences, Royal Irish Academy, Accademia Nazionale dei Lincei (Italy), Academy of Sciences Malaysia, Academy Council of the Royal Society of New Zealand, Royal Swedish Academy of Sciences, Turkish Academy of Sciences, and Royal Society (UK).

2005

http://nationalacademies.org/onpi/06072005.pdf

Eleven national science academies, including all of the largest emitters of greenhouse gases, signed a statement that the scientific understanding of climate change was sufficiently strong to justify prompt action. The statement explicitly endorsed the IPCC consensus and stated:

"...there is now strong evidence that significant global warming is occurring. The evidence comes from direct measurements of rising surface air temperatures and subsurface ocean temperatures and from phenomena such as increases in average global sea levels, retreating glaciers, and changes to many physical and biological systems. It is likely that most of the warming in recent decades can be attributed to human activities (IPCC 2001). This warming has already led to changes in the Earth's climate."

The statement was signed by the science academies of: Brazil, Canada, China, France, Germany, India, Italy, Japan, Russia, the United Kingdom, and the United States.

2007

http://www.pik-potsdam.de/aktuelles/nachrichten/dateien/G8_Academies%20Declaration.pdf

In 2007, seventeen national academies issued a joint declaration reconfirming previous statements and strengthening language based on new research from the fourth assessment report of the IPCC, including the following:

"It is unequivocal that the climate is changing, and it is very likely that this is predominantly caused by the increasing human interference with the atmosphere. These changes will transform the environmental conditions on Earth unless counter-measures are taken."

The thirteen signatories were the national science academies of Brazil, Canada, China, France, Germany, Italy, India, Japan, Mexico, Russia, South Africa, the United Kingdom, and the United States.

2007

http://www.interacademies.net/File.aspx?id=4825

In 2007, the Network of African Science Academies submitted a joint "statement on sustainability, energy efficiency, and climate change:"

"A consensus, based on current evidence, now exists within the global scientific community that human activities are the main source of climate change and that the burning of fossil fuels is largely responsible for driving this change. The Intergovernmental Panel on Climate

Change (IPCC) reached this conclusion with "90 percent certainty" in its Fourth Assessment issued earlier this year. The IPCC should be congratulated for the contribution it has made to public understanding of the nexus that exists between energy, climate and sustainability."

The thirteen signatories were the science academies of Cameroon, Ghana, Kenya, Madagascar, Nigeria, Senegal, South Africa, Sudan, Tanzania, Uganda, Zambia, Zimbabwe, as well as the African Academy of Sciences.

June 2008

http://www.nationalacademies.org/includes/climatechangestatement.pdf

In 2008, the thirteen signers of the 2007 joint academies declaration issued a statement reiterating previous statements and reaffirming "that climate change is happening and that anthropogenic warming is influencing many physical and biological systems." Among other actions, the declaration urges all nations to "(t)ake appropriate economic and policy measures to accelerate transition to a low carbon society and to encourage and effect changes in individual and national behaviour."

The thirteen signatories were the national science academies of Brazil, Canada, China, France, Germany, Italy, India, Japan, Mexico, Russia, South Africa, the United Kingdom, and the United States.

May 2009

In May 2009, thirteen national academies issued a joint statement that said among other things:

"The IPCC 2007 Fourth Assessment of climate change science concluded that large reductions in the emissions of greenhouse gases, principally CO2, are needed soon to slow the increase of atmospheric concentrations, and avoid reaching unacceptable levels. However, climate change is happening even faster than previously estimated; global CO2 emissions since 2000 have been higher than even the highest predictions, Arctic sea ice has been melting at rates much faster than predicted, and the rise in the sea level has become more rapid. Feedbacks in the climate system might lead to much more rapid climate changes. The need for urgent action to address climate change is now indisputable."

The thirteen signatories were the national science academies of Brazil, Canada, China, France, Germany, Italy, India, Japan, Mexico, Russia, South Africa, the United Kingdom, and the United States.

2014

In addition to the statement signed in 2001 by the Royal Flemish Academy of Belgium for Sciences and the Arts, the Academie Royale des Sciences, des Lettres & des Beaux-arts de Belgique (the French language academy in Belgium) issued a formal statement:

http://dailyscience.be/wordpress/wp-content/uploads/2014/12/Position-Ac...

July 2015

https://royalsociety.org/~/media/policy/Publications/2015/21-07-15-climate-communique.PDF

In July 2015, the Royal Society and member organizations issued a joint "U.K. Science Communiqué on Climate Change." In part, that statement reads:

"The scientific evidence is now overwhelming that the climate is warming and that human activity is largely responsible for this change through emissions of greenhouse gases.

Governments will meet in Paris in November and December this year to negotiate a legally binding and universal agreement on tackling climate change.

Any international policy response to climate change must be rooted in the latest scientific evidence. This indicates that if we are to have a reasonable chance of limiting global warming in this century to 2°C relative to the pre-industrial period, we must transition to a zero-carbon world by early in the second half of the century.

To achieve this transition, governments should demonstrate leadership by recognising the risks climate change poses, embracing appropriate policy and technological responses, and seizing the opportunities of low-carbon and climate-resilient growth."

It was signed by: The Academy of Medical Sciences (UK), The Academy of Social Sciences (UK), The British Academy for the Humanities and Social Sciences, The British Ecological Society, The Geological Society (UK), The Challenger Society for Marine Sciences, The Institution of Civil Engineers (UK), The Institution of Chemical Engineers, The Institution of Environmental Sciences, The Institute of Physics, The Learned Society of Wales, London Mathematical Society, Royal Astronomical Society, Royal Economic Society, Royal Geographic Society, Royal Meteorological Society, Royal Society, Royal Society of Biology, Royal Society of Chemistry, Royal Society of Edinburgh, Society for General Microbiology, Wellcome Trust, Zoological Society of London

National Research Council (U.S.)

2010 (one of many statements)

https://www.nap.edu/catalog/12782/advancing-the-science-of-climate-change

Climate change is occurring, is caused largely by human activities, and poses significant risks for -- and in many cases is already affecting -- a broad range of human and natural systems. The compelling case for these conclusions is provided in *Advancing the Science of Climate Change*, part of a congressionally requested suite of studies known as America's Climate Choices. While noting that there is always more to learn and that the scientific process is never closed, the book shows that hypotheses about climate change are supported by multiple lines of evidence and have stood firm in the face of serious debate and careful evaluation of alternative explanations.

[The U.S. National Academies of Sciences have also signed a long series of statements with other national academies around the world in support of the state-of-the-science.]

Natural Science Collections Alliance

The NSCA is a signatory to the June 28, 2016 letter to the U.S. Congress: https://www.eurekalert.org/images/2016climateletter6-28-16.pdf

National Society of Professional Engineers

July 2010

https://www.nspe.org/resources/issues-and-advocacy/take-action/position-statements/air-pollution

Acid rain, toxic air pollutants, and greenhouse gas emissions are a major threat to human health and welfare, as well as plant and animal life. Based on recognized adequate research of the causes and effects of the various forms of air pollution, the federal government should establish environmentally and economically sound standards for the reduction and control of these emissions.

Organization of Biological Field Stations

The OBFS is a signatory to the June 28, 2016 letter to the U.S. Congress: https://www.eurekalert.org/images/2016climateletter6-28-16.pdf

Public Health Institute

The PHI is a signatory to the April 2016 declaration: http://www.lung.org/our-initiatives/healthy-air/outdoor/climate-change/declaration-on-climate-change.html?referrer=https://www.google.com/

Royal Astronomical Society

The RAS is a signatory to the July 21, 2015 UK science communiqué on climate change. https://royalsociety.org/~/media/policy/Publications/2015/21-07-15-climate-communique.PDF

Royal Economic Society

The RES is a signatory to the July 21, 2015 UK science communiqué on climate change. https://royalsociety.org/~/media/policy/Publications/2015/21-07-15-climate-communique.PDF

Royal Geographic Society

The RGS is a signatory to the July 21, 2015 UK science communiqué on climate change. https://royalsociety.org/~/media/policy/Publications/2015/21-07-15-climate-communique.PDF

Royal Meteorological Society 2007, 2011, 2015

https://web.archive.org/web/20110321102247/http://www.rmets.org/news/detail.php?ID=332

https://web.archive.org/web/20070415000000*/http://www.rmets.org/pdf/ipcc.pdf

The Fourth Assessment Report (AR4) of the Inter-Governmental Panel on Climate Change (IPCC) is unequivocal in its conclusion that climate change is happening and that humans are contributing significantly to these changes. The evidence, from not just one source but a number of different measurements, is now far greater and the tools we have to model climate change contain much more of our scientific knowledge within them. The world's best climate scientists are telling us it's time to do something about it.

Carbon Dioxide is such an important greenhouse gas because there is an increasing amount of it in the atmosphere from the burning of fossil fuels and it stays in the atmosphere for such a long time; a hundred years or so. The changes we are seeing now in our climate are the result of emissions since industrialisation and we have already set in motion the next 50 years of global warming - what we do from now on will determine how worse it will get.

The RMS is also a signatory to the July 21, 2015 UK science communiqué on climate change. https://royalsociety.org/~/media/policy/Publications/2015/21-07-15-climate-communique.PDF

Royal Society (U.K.)

The RS is a signatory to the July 21, 2015 UK science communiqué on climate change. https://royalsociety.org/~/media/policy/Publications/2015/21-07-15-climate-communique.PDF

Climate change is one of the defining issues of our time. It is now more certain than ever, based on many lines of evidence, that humans are changing Earth's climate. The atmosphere and oceans have warmed, accompanied by sea-level rise, a strong decline in Arctic sea ice, and other climate-related changes. The evidence is clear.

Royal Society of Biology (Formerly the Institute of Biology)

https://www.rsb.org.uk/policy/policy-issues/environmental-sciences/climate-change/climate-change-statement

We strongly support the introduction of policies to significantly reduce UK and global greenhouse gas emissions, as we feel that the consequences of climate change will be severe.

We believe that biologists have a crucial role to play in developing innovative biotechnologies to generate more efficient and environmentally sustainable biofuels, and to capture and store greenhouse gases from power stations and the atmosphere.

It is important for the government to continue to consult scientists, to review policy, and to encourage new technologies so as to ensure the best possible strategies are used to combat this complex issue.

We are in favour of reducing energy demands, in particular by improvements in public transport and domestic appliances.

As some degree of climate change is inevitable, we encourage the development of adaptation strategies to reduce the effects of global warming on our environment.

Current

https://www.rsb.org.uk/policy/policy-issues/environmental-sciences/climate-change

There is an overwhelming scientific consensus worldwide, and a broad political consensus, that greenhouse gas emissions are affecting global climate, and that measures are needed to reduce these emissions significantly so as to limit the extent of climate change. The term 'climate change' is used predominantly to refer to global warming and its consequences, and this policy briefing will address these issues.

What is global warming?

Although long-term fluctuations in global temperature occur due to various factors such as solar activity, there is scientific agreement that the rapid global warming that has occurred in recent years is mostly anthropogenic, i.e. due to human activity. The absorption and emission of solar radiation by greenhouse gases causes the atmosphere to warm.

What global warming has occurred?

Human activities such as fossil fuel consumption and deforestation have elevated atmospheric levels of greenhouse gases such as carbon dioxide, methane and nitrous oxide significantly since pre-industrial times.

The RSB is also a signatory to the July 21, 2015 UK science communiqué on climate change. https://royalsociety.org/~/media/policy/Publications/2015/21-07-15-climate-communique.PDF

Royal Society of Chemistry

The RSC is a signatory to the July 21, 2015 UK science communiqué on climate change. https://royalsociety.org/~/media/policy/Publications/2015/21-07-15-climate-communique.PDF

Royal Society of Edinburgh

The RSE is a signatory to the July 21, 2015 UK science communiqué on climate change. https://royalsociety.org/~/media/policy/Publications/2015/21-07-15-climate-communique.PDF

Royal Society of New Zealand 2016

http://royalsociety.org.nz/expert-advice/papers/yr2016/climate-change-implications-for-new-zealand/

Key aspects of global climate change

Warming of the climate system is unequivocal, and since the 1950s, many of the observed changes are unprecedented over decades to millennia. The atmosphere and oceans have warmed, the amounts of snow and ice have diminished, and sea level has risen.

Key findings

Global surface temperatures have warmed, on average, by around one degree Celsius since the late 19th century. Much of the warming, especially since the 1950s, is very likely a result of increased amounts of greenhouse gases in the atmosphere, resulting from human activity.

The Northern Hemisphere have warmed much faster than the global average, while the southern oceans south of New Zealand latitudes have warmed more slowly. Generally, continental regions have warmed more than the ocean surface at the same latitudes.

Global sea levels have risen around 19 cm since the start of the 20th century, and are almost certain to rise at a faster rate in future.

Surface temperature is projected to rise over the 21st century under all assessed emission scenarios. It is very likely that heat waves will occur more often and last longer, and that extreme precipitation events will become more intense and frequent in many regions. The ocean will continue to warm and acidify, and global mean sea level will continue to rise.

Relatively small changes in average climate can have a big effect on the frequency of occurrence or likelihood of extreme events.

How the future plays out depends critically on the emissions of greenhouses gases that enter the atmosphere over coming decades.

New Zealand is being affected by climate change and impacts are set to increase in magnitude and extent over time.

Floods, storms, droughts and fires will become more frequent unless significant action is taken to reduce global emissions of greenhouse gases, which are changing the climate.

Even small changes in average climate conditions are likely to lead to large changes in the frequency of occurrence of extreme events. Our societies are not designed to cope with such rapid changes.

Society for General Microbiology

The SGM is a signatory to the July 21, 2015 UK science communiqué on climate change. https://royalsociety.org/~/media/policy/Publications/2015/21-07-15-climate-communique.PDF

Society for Industrial and Applied Mathematics

The SIAM is a signatory to the June 28, 2016 letter to the U.S. Congress: https://www.eurekalert.org/images/2016climateletter6-28-16.pdf

Society for Mathematical Biology

The SMB is a signatory to the June 28, 2016 letter to the U.S. Congress: https://www.eurekalert.org/images/2016climateletter6-28-16.pdf

Society for the Study of Amphibians and Reptiles

The SSAR is a signatory to the June 28, 2016 letter to the U.S. Congress: https://www.eurekalert.org/images/2016climateletter6-28-16.pdf

Society of American Foresters
Adopted December 8, 2008; Revised December 7, 2014

http://www.eforester.org/Main/Issues_and_Advocacy/Statements/Forest_Management_and_Climate_Change.aspx

The Society of American Foresters (SAF) believes that climate change policies and actions should recognize the role that forests play in reducing greenhouse gas (GHG) emissions through 1) the substitution of wood products for nonrenewable building materials, 2) forest biomass substitution for fossil fuel-based energy sources, 3) reducing wildfire and other disturbance emissions, and 4) avoided land-use change. SAF also believes that sustainably managed forests can reduce GHG concentrations by sequestering atmospheric carbon in trees and soil, and by storing carbon in wood products made from the harvested trees. Finally, climate change policies can invest in sustainable forest management to achieve these benefits, and respond to the challenges and opportunities that a changing climate poses for forests.

Of the many ways to reduce GHG emissions and atmospheric particulate pollution, the most familiar are increasing energy efficiency and conservation, and using renewable energy sources as a substitution for fossil fuels. Equally important is using forests to address climate change.

Forests play an essential role controlling GHG emissions and atmospheric GHGs, while simultaneously providing essential environmental and social benefits, including clean water, wildlife habitat, recreation, and forest products that, in turn, store carbon.

Finally, changes in long-term patterns of temperature and precipitation have the potential to dramatically affect forests nationwide through a variety of changes to growth and mortality (USDA Forest Service 2012). Many such changes are already evident, such as longer growing and wildfire seasons, increased incidence of pest and disease, and climate-related mortality of specific species (Westerling et al. 2006). These changes have been associated with increasing concentrations of atmospheric carbon dioxide (CO2) and other GHGs in the atmosphere. Successfully achieving the benefits forests can provide for addressing climate change will therefore require explicit and long-term policies and investment in managing these changes, as well as helping private landowners and public agencies understand the technologies and practices that can be used to respond to changing climate conditions...

Society of Nematologists

The SoN is a signatory to the June 28, 2016 letter to the U.S. Congress: https://www.eurekalert.org/images/2016climateletter6-28-16.pdf

Society of Systematic Biologists

The SSB is a signatory to the June 28, 2016 letter to the U.S. Congress: https://www.eurekalert.org/images/2016climateletter6-28-16.pdf

Soil Science Society of America May 2011, 2016

https://www.soils.org/files/science-policy/asa-cssa-sssa-climate-change-policy-statement.pdf

A comprehensive body of scientific evidence indicates beyond reasonable doubt that global climate change is now occurring and that its manifestations threaten the stability of societies as well as natural and managed ecosystems. Increases in ambient temperatures and changes in related processes are directly linked to rising anthropogenic greenhouse gas (GHG) concentrations in the atmosphere. The potential related impacts of climate change on the ability of agricultural systems, which include soil and water resources, to provide food, feed, fiber, and fuel, and maintenance of ecosystem services (e.g., water supply and habitat for crop landraces, wild relatives, and pollinators) as well as the integrity of the environment, are major concerns.

Around the world and in the United States (US), agriculture—which is comprised of field, vegetable, and tree crops, as well as livestock production—constitutes a major land use which influences global ecosystems. Globally, crop production occupies approximately 1.8 Billion (B) hectares out of a total terrestrial land surface of about 13.5 B hectares. In addition, animal production utilizes grasslands, rangelands, and savannas, which altogether cover about a quarter of the Earth's land. Even in 2010, agriculture remains the most basic and common human occupation on the planet and a major contributor to human well-being. Changes in climate are already affecting the sustainability of agricultural systems and disrupting production.

[The May 2011 Statement was also signed by the American Society of Agronomy and the Crop Science Society of America.]

[The SSSA is also a signatory to the June 28, 2016 letter to the U.S. Congress: https://www.eurekalert.org/images/2016climateletter6-28-16.pdf]

The Academy of Medical Sciences (UK)

The AMS is a signatory to the July 21, 2015 UK science communiqué on climate change. https://royalsociety.org/~/media/policy/Publications/2015/21-07-15-climate-communique.PDF

The Academy of Social Sciences (UK)

The AoSS is a signatory to the July 21, 2015 UK science communiqué on climate change. https://royalsociety.org/~/media/policy/Publications/2015/21-07-15-climate-communique.PDF

The British Academy for the Humanities and Social Sciences

The BAHSS is a signatory to the July 21, 2015 UK science communiqué on climate change. https://royalsociety.org/~/media/policy/Publications/2015/21-07-15-climate-communique.PDF

The British Ecological Society

The BES is a signatory to the July 21, 2015 UK science communiqué on climate change. https://royalsociety.org/~/media/policy/Publications/2015/21-07-15-climate-communique.PDF

The Challenger Society for Marine Sciences

The CSMS is a signatory to the July 21, 2015 UK science communiqué on climate change. https://royalsociety.org/~/media/policy/Publications/2015/21-07-15-climate-communique.PDF

The Geological Society (UK)
November 2010 (updated 2013 and 2015)

https://www.geolsoc.org.uk/~/media/shared/documents/policy/Climate%20Change%20Statement%20final%20%20%20new%20format.pdf?la=en

https://www.geolsoc.org.uk/climaterecord

The last century has seen a rapidly growing global population and much more intensive use of resources, leading to greatly increased emissions of gases, such as carbon dioxide and methane, from the burning of fossil fuels (oil, gas and coal), and from agriculture, cement production and deforestation. Evidence from the geological record is consistent with the physics that shows that adding large amounts of carbon dioxide to the atmosphere warms the world and may lead to: higher sea levels and flooding of low-lying coasts; greatly changed patterns of rainfall; increased acidity of the oceans; and decreased oxygen levels in seawater…

There is now widespread concern that the Earth's climate will warm further, not only because of the lingering effects of the added carbon already in the system, but also because of further additions as human population continues to grow…

[The GS is also a signatory to the July 21, 2015 UK science communiqué on climate change. https://royalsociety.org/~/media/policy/Publications/2015/21-07-15-climate-communique.PDF]

The Institute of Physics

The IoP is a signatory to the July 21, 2015 UK science communiqué on climate change. https://royalsociety.org/~/media/policy/Publications/2015/21-07-15-climate-communique.PDF

The Institution of Chemical Engineers

The ICE is a signatory to the July 21, 2015 UK science communiqué on climate change. https://royalsociety.org/~/media/policy/Publications/2015/21-07-15-climate-communique.PDF

The Institution of Civil Engineers

The ICE is a signatory to the July 21, 2015 UK science communiqué on climate change. https://royalsociety.org/~/media/policy/Publications/2015/21-07-15-climate-communique.PDF

The Institution of Environmental Sciences

The IES is a signatory to the July 21, 2015 UK science communiqué on climate change. https://royalsociety.org/~/media/policy/Publications/2015/21-07-15-climate-communique.PDF

The Learned Society of Wales

The LSoW is a signatory to the July 21, 2015 UK science communiqué on climate change. https://royalsociety.org/~/media/policy/Publications/2015/21-07-15-climate-communique.PDF

The Wildlife Society (international)
November 2011 Statement

http://wildlife.org/wp-content/uploads/2014/05/PS_GlobalClimateChange.pdf

Human activities over the past 100 years have caused significant changes in the earth's climatic conditions, resulting in severe alterations in regional temperature and precipitation patterns that are expected to continue and become amplified over the next 100 years or more. Although climates have varied since the earth was formed, few scientists question the role of humans in exacerbating recent climate change through the increase in emissions of greenhouse gases (e.g., carbon dioxide, methane, water vapor). Human activities contributing to climate warming include the burning of fossil fuels, slash and burn agriculture, methane production from animal husbandry practices, and land-use changes. The critical issue is no longer "whether" climate change is occurring, but rather how to address its effects on wildlife and wildlife-habitats...

Trust For America's Health

The TFAA is a signatory to the April 2016 statement: http://www.lung.org/our-initiatives/healthy-air/outdoor/climate-change/declaration-on-climate-change.html?referrer=https://www.google.com/

The United States of America Nationally Determined Contribution Reducing Greenhouse Gases in the United States: A 2030 Emissions Target. Retrieved from: https://unfccc.int/sites/default/files/NDC/2022-06/United%20States%20NDC%20April%2021%202021%20Final.pdf

U.S. Climate and Health Alliance

The USCHA is a signatory to the April 2016 statement: http://www.lung.org/our-initiatives/healthy-air/outdoor/climate-change/declaration-on-climate-change.html?referrer=https://www.google.com/

University Corporation for Atmospheric Research

The UCAR is a signatory to the June 28, 2016 letter to the U.S. Congress: https://www.eurekalert.org/images/2016climateletter6-28-16.pdf

Wellcome Trust

Wellcome is a signatory to the July 21, 2015 UK science communiqué on climate change. https://royalsociety.org/~/media/policy/Publications/2015/21-07-15-climate-communique.PDF

World Federation of Engineering Organizations May 24, 2016

http://www.wfeo.org/wp-content/uploads/declarations/WFEO_Statement_for_UNFCCC_SB44_Bonn_Meeting.pdf

Now that the world has negotiated the Paris agreement to mitigate GHGs and pursue adaptation to the changing climate, the focus must now turn towards implementation to turn the words into action. The world's engineers are a human resource that must be tapped to contribute to this implementation. All countries use engineers to deliver services that provide the quality of life that society enjoys, in particular, potable water, sanitation, shelter, buildings, roads, bridges, power, energy and other types of infrastructure. There are opportunities to achieve GHG reduction as well as improving the climate resilience of this infrastructure through design, construction and operation all of which require the expertise and experience of engineers. Engineers are problem-solvers and seek to develop feasible solutions that are cost-effective and sustainable.

Engineers serve the public interest and offer objective, unbiased review and advice. Having their expertise to evaluate the technical feasibility and economic viability of proposals to reduce GHGs and to adapt to climate change impacts should be pursued. Engineers input and action is required to implement solutions at country and local levels.

The international organization known as the World Federation of Engineering Organizations consist of members of national engineering organizations from over 90 developing and developed countries representing more than 20 million engineers. The WFEO offers to facilitate contact and engagement with these organizations to identify subject matter experts that will contribute their time and expertise as members of the engineering profession. The expertise of the world's engineers is needed to help successfully implement the Paris agreement. We encourage all countries to engage their engineers in this effort. The WFEO is prepared to assist in this effort.

December 8, 2015

http://www.wfeo.org/wp-content/uploads/declarations/WFEO-COP-21_Engineering_Summit_Statement.pdf

The WFEO consists of national members representing more than 85 countries as well as 10 regional engineering organizations. These members collectively engage with more than 20 million engineers worldwide who are committed to serve the public interest through Codes of Practice and a Code of Ethics that emphasize professional practice in sustainable development, environmental stewardship and climate change.

WFEO, the International Council for Science (ICSU) and the International Social Science Council (ISSC) are co-organizing partners of the UN Major Group on Scientific and Technological Communities, one of the nine major groups of civil society recognized by the United Nations.

Engineers acknowledge that climate change is underway and that sustained efforts must be undertaken to address this worldwide challenge to society, our quality of life and prosperity. Urgent actions are required and the engineering profession is prepared to do its part towards implementing

cost-effective, feasible and sustainable solutions working in partnership with stakeholders.

World Federation of Public Health Associations May 14, 2001

https://web.archive.org/web/20081217173936/http://www.wfpha.org/Archives/01.22%20Global%20Climate%20Change.pdf

Noting the conclusions of the United Nations' Intergovernmental Panel on Climate Change (IPCC) and other climatologists that anthropogenic greenhouse gases, which contribute to global climate change, have substantially increased in atmospheric concentration beyond natural processes and have increased by 28 percent since the industrial revolution….Realizing that subsequent health effects from such perturbations in the climate system would likely include an increase in: heat-related mortality and morbidity; vector-borne infectious diseases,... water-borne diseases...(and) malnutrition from threatened agriculture....the World Federation of Public Health Associations...recommends precautionary primary preventive measures to avert climate change, including reduction of greenhouse gas emissions and preservation of greenhouse gas sinks through appropriate energy and land use policies, in view of the scale of potential health impacts...

World Health Organization June 2016

http://www.who.int/mediacentre/factsheets/fs266/en/

Over the last 50 years, human activities – particularly the burning of fossil fuels – have released sufficient quantities of carbon dioxide and other greenhouse gases to trap additional heat in the lower atmosphere and affect the global climate.

In the last 130 years, the world has warmed by approximately 0.85oC. Each of the last 3 decades has been successively warmer than any preceding decade since 1850.

Sea levels are rising, glaciers are melting and precipitation patterns are changing. Extreme weather events are becoming more intense and frequent...

Many policies and individual choices have the potential to reduce greenhouse gas emissions and produce major health co-benefits. For example, cleaner energy systems, and promoting the safe use of public transportation and active movement – such as cycling or walking as alternatives to using private vehicles – could reduce carbon emissions, and cut the burden of household air pollution, which causes some 4.3 million deaths per year, and ambient air pollution, which causes about 3 million deaths every year.

In 2015, the WHO Executive Board endorsed a new work plan on climate change and health. This includes:

Partnerships: to coordinate with partner agencies within the UN system, and ensure that health is properly represented in the climate change agenda.

Awareness raising: to provide and disseminate information on the threats that climate change presents to human health, and opportunities to promote health while cutting carbon emissions.

Science and evidence: to coordinate reviews of the scientific evidence on the links between climate change and health, and develop a global research agenda.

Support for implementation of the public health response to climate change: to assist countries to build capacity to reduce health vulnerability to climate change, and promote health while reducing carbon emissions.

WHO Call For Urgent Action 2015

http://www.who.int/globalchange/global-campaign/cop21/en/

Climate change is the greatest threat to global health in the 21st century.

Health professionals have a duty of care to current and future generations. You are on the front line in protecting people from climate impacts - from more heat-waves and other extreme weather events; from outbreaks of infectious diseases such as malaria, dengue and cholera; from the effects of malnutrition; as well as treating people that are affected by cancer, respiratory, cardiovascular and other non-communicable diseases caused by environmental pollution.

Already the hottest year on record, 2015 will see nations attempt to reach a global agreement to address climate change at the United Nations Climate Change Conference (COP) in Paris in December. This may be the most important health agreement of the century: an opportunity not only to reduce climate change and its consequences, but to promote actions that can yield large and immediate health benefits, and reduce costs to health systems and communities…

World Meteorological Organization 2016

https://public.wmo.int/en/our-mandate/climate

Since the beginning of the 20th century, scientists have been observing a change in the climate that cannot be attributed solely to natural influences. This change has occurred faster than any other climate change in Earth's history and will have consequences for future generations. Scientists agree that this climate change is anthropogenic (human-induced). It is principally attributable to the increase of certain heat absorbing greenhouse gases in our atmosphere since the industrial revolution. The ever-increasing amount of these gases has directly lead to more heat being retained in the atmosphere and thus to increasing global average surface temperatures. The partners in the WMO Global Atmosphere Watch (GAW) compile reliable scientific data and information on the chemical composition of the atmosphere and its natural and anthropogenic change. This helps to improve the understanding of interactions between the atmosphere, the oceans and the biosphere.

November 8, 2016

https://public.wmo.int/en/media/press-release/global-climate-2011-2015-hot-and-wild

The World Meteorological Organization has published a detailed analysis of the global climate 2011-2015 – the hottest five-year period on record - and the increasingly visible human footprint on extreme weather and climate events with dangerous and costly impacts.

The record temperatures were accompanied by rising sea levels and declines in Arctic sea-ice extent, continental glaciers and northern hemisphere snow cover.

All these climate change indicators confirmed the long-term warming trend caused by greenhouse gases. Carbon dioxide reached the significant milestone of 400 parts per million in the atmosphere for the first time in 2015, according to the WMO report which was submitted to U.N. climate change conference.

Zoological Society of London

The Zoological Society is a signatory to the July 21, 2015 UK science communiqué on climate change. https://royalsociety.org/~/media/policy/Publications/2015/21-07-15-climate-communique.PDF

Edited, compiled by Dr. Peter Gleick

Gleick, P. (2017). Statements on Climate Change from Major Scientific Academies, Societies, and Associations (January 2017 update). Science Blogs. Retrieved from: https://scienceblogs.com/significantfigures/index.php/2017/01/07/statements-on-climate- change-from-major-scientific-academies-societies-and-associations-january-2017-update

Appendix II

The Paris Agreement (Highlights)

The Paris Agreement requires all Parties to put forward their best efforts through "nationally determined contributions" (NDCs) and to strengthen these efforts in the years ahead. This includes requirements that all Parties report regularly on their emissions and on their implementation efforts. There will also be a global stock take every 5 years to assess the collective progress towards achieving the purpose of the agreement and to inform further individual actions by Parties.

The Paris Agreement addresses crucial areas necessary to combat climate change. Some of the key aspects of the Agreement are set forth below:

- ***Long-term temperature goal*** (Art. 2) – The Paris Agreement, in seeking to strengthen the global response to climate change, reaffirms the goal of limiting global temperature increase to well below 2 degrees Celsius, while pursuing efforts to limit the increase to 1.5 degrees.
- ***Global peaking* and ‹climate neutrality›** (Art. 4) –To achieve this temperature goal, Parties aim to reach global peaking of greenhouse gas emissions (GHGs) as soon as possible, recognizing peaking will take longer for developing country Parties, so as to achieve a balance between anthropogenic emissions by sources and removals by sinks of GHGs in the second half of the century.

- ***Mitigation*** (Art. 4) – The Paris Agreement establishes binding commitments by all Parties to prepare, communicate and maintain a nationally determined contribution (NDC) and to pursue domestic measures to achieve them. It also prescribes that Parties shall communicate their NDCs every 5 years and provide information necessary for clarity and transparency. To set a firm foundation for higher ambition, each successive NDC will represent a progression beyond the previous one and reflect the highest possible ambition. Developed countries should continue to take the lead by undertaking absolute economy-wide reduction targets, while developing countries should continue enhancing their mitigation efforts, and are encouraged to move toward economy-wide targets over time in the light of different national circumstances.
- ***Sinks and reservoirs*** (Art.5) –The Paris Agreement also encourages Parties to conserve and enhance, as appropriate, sinks and reservoirs of GHGs as referred to in Article 4, paragraph 1(d) of the Convention, including forests.
- ***Voluntary cooperation/Market- and non-market-based approaches*** (Art. 6) – The Paris Agreement recognizes the possibility of voluntary cooperation among Parties to allow for higher ambition and sets out principles – including environmental integrity, transparency and robust accounting – for any cooperation that involves internationally transferal of mitigation outcomes. It establishes a mechanism to contribute to the mitigation of GHG emissions and support sustainable development, and defines a framework for non-market approaches to sustainable development.
- ***Adaptation*** (Art. 7) – The Paris Agreement establishes a global goal on adaptation – of enhancing adaptive capacity, strengthening resilience and reducing vulnerability to climate change in the context of the temperature goal of the Agreement.

It aims to significantly strengthen national adaptation efforts, including through support and international cooperation. It recognizes that adaptation is a global challenge faced by all. All Parties should engage in adaptation, including by formulating and implementing National Adaptation Plans, and should submit and periodically update an adaptation communication describing their priorities, needs, plans and actions. The adaptation efforts of developing countries should be recognized

- ***Loss and damage*** (Art. 8) – The Paris Agreement recognizes the importance of averting, minimizing and addressing loss and damage associated with the adverse effects of climate change, including extreme weather events and slow onset events, and the role of sustainable development in reducing the risk of loss and damage. Parties are to enhance understanding, action and support, including through the Warsaw International Mechanism, on a cooperative and facilitative basis with respect to loss and damage associated with the adverse effects of climate change.
- ***Finance, technology and capacity-building support*** (Art. 9, 10 and 11) – The Paris Agreement reaffirms the obligations of developed countries to support the efforts of developing country Parties to build clean, climate-resilient futures, while for the first time encouraging voluntary contributions by other Parties. Provision of resources should also aim to achieve a balance between adaptation and mitigation. In addition to reporting on finance already provided, developed country Parties commit to submit indicative information on future support every two years, including projected levels of public finance. The agreement also provides that the Financial Mechanism of the Convention, including the Green Climate Fund (GCF), shall serve the Agreement. International cooperation on climate-safe technology development and transfer and building capacity

in the developing world are also strengthened: a technology framework is established under the Agreement and capacity-building activities will be strengthened through, inter alia, enhanced support for capacity building actions in developing country Parties and appropriate institutional arrangements. Climate change education, training as well as public awareness, participation and access to information (Art 12) is also to be enhanced under the Agreement.

- ***Climate change education, training, public awareness, public participation and public access to information*** (Art 12) is also to be enhanced under the Agreement.
- ***Transparency*** (Art. 13), ***implementation and compliance*** (Art. 15) – The Paris Agreement relies on a robust transparency and accounting system to provide clarity on action and support by Parties, with flexibility for their differing capabilities of Parties. In addition to reporting information on mitigation, adaptation and support, the Agreement requires that the information submitted by each Party undergoes international technical expert review. The Agreement also includes a mechanism that will facilitate implementation and promote compliance in a non-adversarial and non-punitive manner, and will report annually to the CMA.
- ***Global Stocktake*** (Art. 14) – A "global stocktake", to take place in 2023 and every 5 years thereafter, will assess collective progress toward achieving the purpose of the Agreement in a comprehensive and facilitative manner. It will be based on the best available science and its long-term global goal. Its outcome will inform Parties in updating and enhancing their actions and support and enhancing international cooperation on climate action.
- ***Decision 1/CP.21*** also sets out a number of measures to enhance action prior to 2020, including strengthening the tech-

nical examination process, enhancement of provision of urgent finance, technology and support and measures to strengthen high-level engagement. For 2018 a facilitative dialogue is envisaged to take stock of collective progress towards the long-term emission reduction goal of Art 4. The decision also welcomes the efforts of all non-Party stakeholders to address and respond to climate change, including those of civil society, the private sector, financial institutions, cities and other subnational authorities.

United Nations (2020). What is the Paris Agreement? *2020 United Nations Framework Convention on Climate Change.* Retrieved from: https://unfccc.int/process-and- meetings/the-paris-agreement/what-is-the-paris-agreement

Appendix III
Frequently Asked Questions

Overarching Frequently Asked Questions taken directly from the IPCC 2022 report on climate change (Working Group, 2022).

"The WGII overarching Frequently Asked Questions (FAQs) are an outreach material. They are based on the WGII Report and aim to help to interpret its concepts and findings to a broad audience. " The original version of the FAQ's can be found at: https://www.ipcc.ch/report/ar6/wg2/about/frequently-asked-questions

FAQ 1: What are the new insights on climate impacts, vulnerability and adaptation from IPCC?

> ***Climate change is affecting nature, people's lives and infrastructure everywhere. Its dangerous and pervasive impacts are increasingly evident in every region of our world. These impacts are hindering efforts to meet basic human needs and they threaten sustainable development across the globe.***
>
> *All life on Earth – from ecosystems to human civilization – is vulnerable to a changing climate. Since the first IPCC reports, the evidence has become stronger: our world is warming and dangerous climate change and extreme events are increasingly impacting nature and people's lives everywhere. This can be seen in the depths of the ocean and at the top of the highest mountains; in rural areas as well as in cities. The extent and magnitude of climate change impacts are larger than estimated in previous assessments. They are causing severe and widespread disruption in nature and in soci-*

ety; reducing our ability to grow nutritious food or provide enough clean drinking water, thus affecting people's health and well-being and damaging livelihoods. In summary, the impacts of climate change are affecting billions of people in many different ways.

Since the Fifth IPCC Assessment Report, published in 2014, a wider range of impacts can be attributed to climate change. In other words: there is new knowledge that climate change caused them or made them more likely. In particular, increasing heat and extreme weather are driving plants and animals on land and in the ocean towards the poles, to higher altitudes, or to the deeper ocean waters. Many species are reaching limits in their ability to adapt to climate change, and those that cannot adjust or move fast enough are at risk of extinction. As a result, the distribution of plants and animals across the globe is changing and the timing of key biological events such as breeding or flowering is altering. These trends are affecting food webs. In many cases, this reduces the ability of nature to provide the essential services that we depend on to survive – such as coastal protection, food supply or climate regulation via carbon uptake and storage.

Changes in temperature, rainfall, and extreme weather have also increased the frequency and spread of diseases in wildlife, agriculture, and people. We see a lengthening wildfire season and increases in the area burned. Roughly half of the world's population currently experiences severe water shortages at some point during the year, in part due to climate change and extreme events such as flooding and droughts. Drought conditions have become more frequent in many regions, negatively affecting agriculture and energy production from hydroelectric power plants.

People living in cities nowadays face higher risks of heat stress, reduced air quality because of wildfire, lack of water, food shortages and other impacts caused by climate change and its effect on supply chains, transport networks and other critical infrastructure. Globally, climate change is increasingly causing injuries, illness, malnutrition, threats to physical and mental health and well-being, and even deaths. It is making hot areas even hotter and drastically reducing the time people can spend outside, which means that some outdoor workers cannot work the required hours and thus will earn less.

Climate change impacts are expected to intensify with additional warming. It is also an established fact that they are interacting with multiple other societal and environmental challenges. These include a growing world population, unsustainable consumption, a rapidly increasing number of people living in cities, significant inequality, continuing poverty, land degradation, biodiversity loss due to land-use change, ocean pollution, overfishing and habitat destruction as well as a global pandemic. Where trends intersect they can reinforce each other, intensifying risks and impacts, which affect the poor and most vulnerable people the hardest.

Climate change risks and impacts can be reduced, within limits, if humans and nature adapt to the changing conditions. The scale and scope of actions to reduce climate risks (adaptation) have increased worldwide. Individuals and households along with communities, businesses, religious groups and social movements are adapting to climate change already. However, the Working Group II Report identifies large gaps between ongoing efforts, and adaptation needed to cope with current levels of warming, with the scale of the challenge varying in different regions. The report also highlights that the effectiveness of available adaptation options decreases with every increment of warming. Successful adaptation requires urgent, more ambitious and accelerated action and, at the same time, rapid and deep cuts in greenhouse gas emissions. The quicker and further emissions fall, the more scope there is for people and nature to adapt (see FAQ 4 for further details).

The focus of our new report is on solutions. It highlights the importance of fundamental changes in society at the same time as conserving, restoring and safeguarding nature in order to meet the Paris Agreement and the Sustainable Development Goals. It is clear now that minor, marginal, reactive or incremental changes won't be sufficient. In addition to technological and economic changes, shifts in most aspects of society are required to overcome limits to adaptation, build resilience, reduce climate risk to tolerable levels, guarantee inclusive, equitable and just development and achieve societal goals without leaving anyone behind.

The strong and interdependent relationships between climate, nature and people are fundamental to reaching these goals. This is emphasized more strongly in the new Working Group II Report than in previous IPCC assessments. We now know that a healthy planet is fundamental to secure a liveable future for people on Earth and that's why we say that the needs of climate, nature and local communities have to be considered together and prioritized in decision making and planning - every day and in every region of our world.

FAQ 2: How will nature and the benefits it provides to people be affected by higher levels of warming?

Healthy ecosystems and rich biodiversity underpin human survival. They provide countless services that make our Earth a habitable place. However, climate change and increases in extreme weather events are drastically and progressively impacting nature, weakening the structure, functioning and resilience of ecosystems. As a result, nature's contributions to human well-being are diminishing, threatening sustainable and just development – now and in the future.

The world's ecosystems on land, in freshwater and in the ocean provide a wide array of essential services to humans. They produce the food we eat and the oxygen we breathe. They filter our water, recycle nutrients and help to limit global warming by storing large amounts of carbon. Furthermore, they cool the air and offer "green" or "blue" spaces such as parks and lakes for fun, adventure and relaxation, thus improving our health and mental well-being. In short, healthy ecosystems are essential for human survival and make our Earth liveable.

Climate change – with its slow-onset events like sea level rise and ocean acidification and increases in extremes – is drastically and progressively affecting our world's biodiversity and ecosystems. Increasing temperatures and extreme events such as droughts, floods and heatwaves are exposing plants and animals to climatic conditions not experienced for at least tens of thousands of years. Observed increases in their frequency and intensity are starting to exceed the ability of many species to adapt. Based on increased observations and a better understanding of processes, we now know that the extent and magnitude of climate change impacts on nature are greater than previously assessed. The impacts we see today are appearing much faster, they are more disruptive and more widespread than we expected 20 years ago. And we know that climate change is strongly adding to, and even amplifying, the other stressors: many of our world's ecosystems are already facing a biodiversity crisis due to human impacts such as deforestation, pollution, overfishing and land-use change. For numerous ecosystems, climate change impacts are an additional stress and even a deadly burden, depending on the level of global warming.

We see a growing number of scientific studies that present multiple lines of evidence showing climate change impacts. Increasing temperatures and extreme events change the seasonal timing of key biological events such as flowering, when animals emerge from hibernation, or annual migration, causing mismatches with important seasonal food sources. Examples include the timing of fish spawning and plankton blooms that fish larvae depend on for food, and insect availability at the time when birds are breeding.

Changing climatic conditions, including warming, also progressively shift plants and animals to higher latitudes, higher elevations or deeper ocean waters. Approximately half of the many thousands of species studied on land and in the ocean already show corresponding responses, leading to climate-caused local population extinctions and shifts in vegetation zones. In the ocean, marine plants and animals including entire communities have shifted their distributions poleward at an average speed of 59 km per decade due to increasing water temperatures. Ocean acidification and decreasing oxygen in the water also play a part. Together all three processes have caused a reorganisation of biodiversity over the past 50 years, especially at the ocean surface. Those species that cannot adjust or move fast enough are at high risk of becoming extinct.

As a result, the geographic patterns and the regional and local abundance of plants and animals are changing, with potentially severe impacts for herders, farmers, fishers, hunters, foragers and other people who directly rely on nature's services. As an example, the sustainable potential for fishery catches of several marine fish and shellfish is estimated to have decreased by 4.1% globally in the 70 years between 1930 and 2010 due to ocean warming. Regions like the North Sea and Celtic Sea have experienced even stronger decreases in fisheries productivity primarily due to warming, but other human activities such as overfishing have played a role as well.

Although there have been some positive impacts on agricultural productivity in some high-latitude regions, with ongoing warming, current global crop and livestock areas will become increasingly unsuitable. Even in a world with low greenhouse gas emissions (warming below 1.6°C by 2100), some 8% of today's farmland is projected to become climatically unsuitable by 2100. Under the same conditions, fisher people in Africa's tropical regions are projected to lose between 3 to 41% of their fisheries' yield by the end of the century due to local extinctions of marine fish. Fisheries provide the main source of protein for about one-third of people living in Africa. It supports the livelihoods of 12.3 million people. Declining fish harvests could leave millions of people vulnerable to malnutrition.

Increases in frequency and severity of extreme weather events such as heatwaves and heavy rain are occurring across all continents and all oceans, resulting in local mass die-offs and local extinctions because the impacts of those events already exceed what many organisms can tolerate. Prominent examples of species being pushed way beyond their temperature limits are reef-building warm-water corals that are dying. Their global decline shows that we don't need to look into the future to recognize the urgency of climate action.

The more often ecosystems are impacted by extreme events and the more intense the event, the further they are pushed towards so-called tipping points. Beyond those points, abrupt and in some cases irreversible changes can occur – such as species going extinct. This risk increases steeply with rises in global temperature. Current projections imply that at a global warming level of 2°C by 2100, up to 18% of all species on land will be at high risk of going extinct. If the world warms up to 4°C, every second plant or animal species that we know of will be threatened.

The extinction risk is especially high for cold-loving species living in the high mountains or in polar regions, where climate change impacts are unfolding at global maximum speed and extent. With global warming of around 4°C by 2100 (very high greenhouse gas emissions scenario), mass mortalities and extinctions are expected that will irreversibly alter globally important areas, including those that host exceptionally rich biodiversity such as tropical coral reefs and cold-water kelp forests and the world's rainforests. Even at lower levels of warming of 2°C or less, polar fauna (including fish, penguins, seals, and polar bears), tropical coral reefs and mangroves will be under serious threat.

Impacts will continue to increase, weakening the structure, functioning and resilience of ecosystems and thus the services they provide, including their ability to regulate our world's climate. At present, ecosystems are removing and storing more carbon from the atmosphere than they emit, hence naturally helping to balance global warming. However, logging in remaining natural forests; draining and burning of peatlands, and increasing climate change impacts, such as the thawing of Arctic permafrost, are causing some of those ecosystems to emit more carbon to the atmosphere (from decomposing dead plant material) than they naturally remove (through vegetation growth). The thawing of Arctic permafrost will cause enhanced methane release with further warming. All of these systems have the potential to contribute to accelerating climate change and further climate change will exacerbate this.

This and other trends can still be reversed by restoring, rebuilding and strengthening ecosystems and by managing them sustainably – which will also support the well-being and livelihoods of people. To achieve this balance, drastic greenhouse gas emissions reduc-

tions are required now to avoid further global warming and its deadly impacts on ecosystems around the world. For indeed, humans are just one of the many living organisms in our beautiful and complex world.

FAQ 3: How will climate change affect the lives of today's children tomorrow, if no immediate action is taken?

Climate change impacts are increasingly being felt in all regions of the world with growing challenges for water availability, food production and the livelihoods of millions of people. We also know that impacts will continue to increase if drastic cuts in greenhouse gas emissions are further delayed – affecting the lives of today's children tomorrow and those of their children much more than ours. But science is also clear: with immediate action now, drastic impacts can still be prevented.

The scientific assessment in the WGII Report addresses the near-term (up to 2040), mid-term (2041-2060) and the long-term (2081-2100). Today, the latter two milestones may seem far away, but children who were born in 2020 will be 20 years old in 2040 and 80 years old in 2100. The end of the century is less than a lifetime away. Actions taken now to reduce emissions of heat-trapping greenhouse gases drastically and adapt to a changing climate will have a profound effect on the quality of their lives and their children's lives, as well as their health, well-being, and security. We also have to take into account that by 2050 almost 70% of the world's growing population will live in urban areas, many in unplanned or informal settlements. As a result, today's children and future generations are more likely to be exposed and vulnerable to climate change and related risks such as flooding, heat stress, water scarcity, poverty, and hunger. Children are amongst those suffering the most, as we see today.

But what is our children's future going to look like, if we do not limit global warming to well below 2°C, preferably to 1.5°C, compared to pre-industrial temperature? Based on the Working Group II assessment, we know that global warming is already changing much of the world as we know it. Its impacts will intensify in the coming decades with profound implications for all aspects of human life around the world. Our food and water supplies, our cities, infrastructure and economies as well as our health and well-being will be affected.

For example: children aged ten or younger in the year 2020 are projected to experience a nearly four-fold increase in extreme events under 1.5°C of global warming by 2100, and a five-fold increase under 3°C warming. Such increases in exposure would not be experienced by a person aged 55 in the year 2020 in their remaining lifetime under any warming scenario.

Globally, the percentage of the population exposed to deadly heat stress is projected to increase from today's 30% to 48-76% by the end of the century, depending on future warming levels and location. If the world warms more than 4°C by 2100, the number of days with climatically stressful conditions for outdoor workers will increase by up to 250 workdays per year by century's end in some parts of South Asia, tropical sub-Saharan Africa and parts of Central and South America. This would cause negative consequences such as reduced food production and higher food prices. In Europe, the number of people at risk of heat stress will increase two- to three-fold at 3°C global warming compared to warming levels of 1.5°C.

With ongoing global warming, today's children in South and Southeast Asia will witness increased losses in coastal settlements and infrastructure due to flooding caused by unavoidable sea level rise, with very high losses in East Asian cities. By mid-century, more than a billion people living in low-lying coastal cities and settlements globally are projected to be at risk from coastal-specific climate hazards. Many of those will be forced to move to higher ground, which will increase competition for land and the probability of conflict and forced relocation.

Climate change will impact water quality and availability for hygiene, food production and ecosystems due to floods and droughts. Globally, 800 million to 3 billion people are projected to experience chronic water scarcity due to droughts at 2°C warming, and up to approximately 4 billion at 4°C warming, considering the effects of climate change alone, with present-day population. Children growing up in South America will face an increasing number of days with water scarcity and restricted water access, especially those living in cities and in rural areas depending on water from glaciers. As the Andean glaciers and snowcaps continue to melt, the amount of available water decreases as the glaciers shrink or disappear entirely. Countries in Central America will experience more frequent and stronger storms or hurricanes and heavy rainfall, causing river flooding.

Today's young people and future generations will also witness stronger negative effects of climate change on food production and availability. The warmer it gets, the more difficult it will become to grow or produce, transport, distribute, buy, and store food – a trend that is projected to hit poor populations the hardest. Depending on future policies and climate and adaptation actions taken, the number of people suffering from hunger in 2050 will range from 8 million to up to 80 million people, with most severely affected populations concentrated in Sub-Saharan Africa, South Asia and Central America. Under a high vulnerability-high warming scenario, up to 183 million additional people are projected to become undernourished in low-income countries due to climate change by 2050.

Africa is the continent with the world's youngest population (40% of the population are under 15 years old). Here, climate change will significantly increase the number of children with severe stunting (impaired growth and development which often leads to limited physical and cognitive potential), by approximately 1.4 million by 2050 under 2.1°C of warming due to malnutrition. The lack of food and under-nutrition are strongly linked with hot climates in the sub-Saharan area and less rainfall in West and Central Africa. Climate change can undermine children's educational attainment, thus reducing their chances for well-paid jobs or higher incomes later in life.

The concerning news is: all these projected impacts will not only reduce the prospects of sustainable development, but our Working Group II Report also projects an increase in poverty and inequality as well as increased involuntary migration of people due to climate change. These responses follow expected climate-driven increases in the frequency and strength of regional wildfires, increased floods and droughts, and an increase in temperature-related incidences of vector-borne, water-borne and food-borne diseases such as dengue, malaria, cholera and Rift Valley Fever.

In addition, we now know that multiple climate hazards will occur simultaneously more often in the future. They may reinforce each other and result in increased impacts and risks to nature and people that are more complex and more difficult to manage. For example, reductions in crop yields due to heat and drought, made worse by reduced productivity because of heat stress among farmworkers, will increase food prices, reduce household incomes and lead to health risks from malnutrition, as well as climate-related deaths, especially in tropical regions.

But there is also positive news: all these risks can be reduced substantially by taking urgent action to limit global warming and by strengthening our adaptation efforts – for example by protecting and conserving nature, and by improving planning and management of our cities (for details see TS FAQ 5). The youth movement, together with many non-governmental organizations, has led to a rising wave of public global awareness of climate change and its life-threatening impacts. To successfully secure our own future and the future of the coming generations, climate risks must be factored into each decision and planning. We have the knowledge and the tools. Now it is our choice to make.

FAQ 4: 4. How are people adapting to the effects of climate change and what are the known limits to adaptation?

If we want to deal with climate hazards and reduce risks for people and ecosystems that come from climate change, we have to adapt. Awareness of climate risks and action to reduce them have increased globally, but progress is uneven and our report highlights large gaps between adaptation action taken and what is needed in many regions. These gaps are caused by for example a lack of funding, political commitment, reliable information and sense of urgency. This leads to the most vulnerable people and ecosystems being hit hardest by climate change. In addition, the report clarifies: adaptation is essential to reduce harm, but if it is to be effective, it must go hand-in-hand with ambitious reductions in greenhouse gas emissions because with increased warming the effectiveness of many adaptation options declines.

Due to climate change, the world is experiencing higher temperatures, rising sea levels, and increased extreme events that impact life on land and in the oceans. To cope with these changes and avoid drastic losses and damages, humans and nature must adapt.

For plants and animals, this means either adjusting to the changing climate and its effects by spending more time and energy on life-sustaining measures (like regulating their body temperature, selecting cooler places or staying hydrated) or, if possible, shifting to regions where environmental conditions are still in the climatic range that organisms are used to. For people and society, adaptation to climate change means adjusting our behaviour (e.g. where we choose to live; the way we plan our cities and settlements) and adapting our infrastructure (e.g. greening of urban areas for water storage) to deal with the changing climate - today and in the future.

Adapting successfully requires an analysis of risks caused by climate change and the implementation of measures in time to reduce these risks. That is the reason why IPCC authors ask five questions when assessing progress in climate adaptation regionally and globally: 1) Is there an awareness that climate change is causing risks? 2) Are the current and future extent of climate risks being assessed? 3) Have adaptation measures to reduce these risks been developed and included in planning? 4) Are those adaptation measures being implemented? 5) Are their implementation and effectiveness in reducing risks monitored and evaluated?

In our Working Group II Report, we conclude that the awareness and assessment of current and future climate risks have increased worldwide. National and local governments as well as corporations and civil society acknowledge the growing need for adaptation. At least 170 countries and many cities now have adaptation included in their climate policies and planning processes. Pilot projects and local experiments are being implemented in different sectors.

However, given the rate and scope of climate change impacts, actions on assessing and communicating risks, as well as on implementing adaptation are insufficient. For instance, current adapta-

tion-related responses across all sectors and regions are dominated by minor modifications to usual practices or measures for dealing with extreme weather events – often allowing small or locally contained reductions of risks only. Whilst this may suffice in the short term, the long-term risks may require more extensive, transformative changes in our behaviour and infrastructure. In brief: ambition, scope, and progress on reducing climate risks are rising, but not by enough. Substantial adaptation gaps still exist, especially among populations with lower income. At the current rate of planning and implementation, these adaptation gaps will continue to grow. According to our new report, the world is currently under-prepared for the coming climate change impacts, particularly beyond 1.5°C global warming.

But there are also still large gaps in our understanding of climate change adaptation. For example, the extent to which adaptation actions are reducing climate risk, and for whom, is not always clear. Another important question is whether adaptation actions may have unintended consequences or side effects, causing more harm than good (this is called maladaptation). Built defences, such as sea walls, might protect coastal areas in the short term but their construction can destroy coastal ecosystems such as coral reefs. In the long term, these defences can even increase risks to people living behind them as more families move to an area that is supposedly safe to live in – as long as the sea wall is not overtopped or destroyed.

In our assessment we show that, in a warming world, measures that are effective now in one place might not work in 20 years, or in other places, which is why the monitoring and evaluation of the implemented actions are so important. Adaptation strategies might have to be revised constantly and those revisions will be most efficient if they are fact- and data-driven. But only a very few nations already have operational frameworks in place to track and evaluate implementation and results.

The Working Group II Report emphasises that the earlier the adaptation measures are implemented, the more the world will benefit because the potential to reduce climate risks through adaptation is higher until mid-century, and for global warming levels below 1.5°C. At higher levels of warming, the effectiveness of most land- and water-based adaptation options starts declining, and the extent of residual risks increases, as do the chances of future unintended consequences.

By investing in adaptation now, the world will avoid higher investments in the future because the potential benefits of adaptation activities outweigh their costs in the long term. In addition, adaptation can generate multiple benefits. Through various adaptation actions we may be able to secure productivity of fisheries, agriculture and companies, foster innovation, health and well-being, strengthen food security and peoples' livelihoods, and rebuild and strengthen nature, while at the same time reducing climate risks and damages.

The world should also be aware that the availability of adaptation options is constrained by limitations faced by the natural world and people, especially at higher degrees of warming. Biophysical, institutional, financial, social and cultural barriers can lead to soft and hard adaptation limits, particularly when combined. Hard limits occur when adaptive actions become infeasible to avoid risks. One prominent example is when small islands become uninhabitable due to sea level rise and lack of sufficient freshwater. In that case, inhabitants may have no other option than to abandon their homes. Soft limits, in contrast, can be overcome if additional financial, institutional or technological support becomes available. With sufficient funding, for instance, cities can invest in new parks and lakes, creating new spaces for citizens to find shade and cool down during heatwaves.

Our report finds that many species and ecosystems are currently near or beyond their hard adaptation limits, and people that rely on them to survive, are currently near or beyond their soft adaptation limits. Californian almonds for instance are predicted to increase their potential geographical range under climate warming, yet a trend of increasing drought has already resulted in trees being removed due to lack of access to irrigation water. This development hits small-scale farmers the hardest.

A lack of political commitment and funding as well as weak institutional capacities limit the implementation of adaptation options in agriculture, fisheries, aquaculture and forestry. In cities, governance capacity, financial support and the legacy of past urban infrastructure investment constrain how cities and settlements are able to adapt. We also see that in cities, the gap between what can be adapted to and what has been implemented is uneven. It is larger for the poorest 20% of the population than for the wealthiest 20%.

Poverty and inequality both present significant adaptation limits, resulting in unavoidable impacts for vulnerable groups, including women, young people, the elderly, ethnic and religious minorities, Indigenous People, and refugees. Climate change is likely to force many of them to switch from agriculture as the main source of income to other forms of wage labour, with implications for labour migration and urbanization.

Climate change is a global threat to which all people and ecosystems are vulnerable. Without effective adaptation, climate change has the potential to reverse the developmental gains in our world and push millions of people further into poverty. To avoid mount-

ing losses, urgent accelerated action is required to adapt to climate change while making rapid, deep cuts in greenhouse gas emissions to limit warming so that we keep the range and scope of adaptation options as wide as possible.

FAQ 5: What strategies could increase the climate resilience of people and nature?

Nature offers a lot of untapped potential, not only to reduce climate risks, and deal with the causes of climate change, but also to improve people's lives. By restoring and safeguarding ecosystems on land and in the ocean, we help plants and animals to build climate resilience. Nature, in turn, can help us regulate the climate, give us clean, safe water, control pests and diseases and pollinate our crops. However, investing in nature alone won't be enough. To secure a healthy, liveable planet for everyone, we need to transform our way of life fundamentally, especially key elements such as our industry and energy sector as well as how cities and infrastructures are planned and built. Taking action now gives us the best chance of success.

Climate change is a threat to human well-being and the health of the planet. According to our new report, it requires urgent and far-reaching actions and fundamental changes in all aspects of human life to increase people's and nature's ability to cope with, and respond to, climate change. One key to success is acknowledging climate, biodiversity, and human society as a coupled system, meaning that all components are interlinked. If we change one of them, it will affect the other two as well. Based on this recognition, conservation and climate change responses would be planned and implemented hand in hand – not only locally, but across landscapes, in cities as well as in rural areas, across sectors, state and country borders. All actions and decisions would be based on the overarching goal to get the best outcomes for climate, biodiversity and the people living in the areas, where actions will be implemented.

While planning those actions, we should keep in mind that only diverse and healthy ecosystems are able to provide the services that are essential for reducing climate change risks. Thus, protecting and restoring ecosystems on land and in the ocean is a key element for success. A range of scientific evidence indicates that the capacity to provide these services relies upon 30 to 50% of Earth's surface (land, freshwater and ocean) to be effectively conserved and for natural resources to be sustainably managed.

An increasing body of evidence demonstrates that climatic risks to people can be lowered by strengthening nature, meaning that we invest in protecting nature and rebuilding ecosystems to benefit both people and biodiversity. Flood risk along rivers, for instance, can be reduced by restoring wetlands and other natural habitats in flood plains, by restoring natural courses of rivers, and by using trees to create shade. Cities can be cooled by parks and ponds and by greening streets and buildings' rooftops and walls. Farmers may increase their businesses' climate resilience by diversifying their crops and livestock, by planting trees and bushes on the fields for shade and organic manure (agroecological farming), by increasing soil health (more soil organic matter), and by combining crops, livestock and natural elements such as trees and bushes.

Actions and solutions that safeguard nature are relatively inexpensive in many parts of the world because they do not rely on complex machinery or on the development of extensive infrastructure. However, to realize potential benefits and avoid harm, it is essential that these solutions are deployed in the right places and with the right approaches for that area, guided by local and indigenous knowledge, scientific understanding and practical expertise. Knowledge is the key.

But relying on nature alone won't be enough. An overall increase in climate resilience requires two combined sets of actions: first, a wide range of actions that reduce human-induced greenhouse gas emissions drastically; secondly, a similarly wide range of actions that transform the way we live our lives and puts human society on the path of sustainable development. The latter is fundamental to enable climate action because, without sufficient knowledge and income as well as governmental support and a fair chance of participation in decision-making processes, many human communities won't be able to contribute to emission reductions or adapt to change. That is especially true for the very poor, for whom struggling to feed the family may occupy all their time and resources.

That is the reason why sustainable development in a climate context includes for example clean energy generation, circular economies, healthy diets from sustainable farming, appropriate urban planning and transport, universal health coverage and social protection, training and education as well as water and energy access for everyone to help to reduce poverty.

The risks posed by climate change vary by location, time, and among different populations. That means each community and each nation has its own starting point for climate adaptation and sustainable development. But, whichever pathway is followed, Climate Resilient Development will only be possible with fundamental changes in five major areas: 1) in our world's energy systems; 2) in the way we use, manage and safeguard the land and freshwater, the oceans and their respective ecosystems; 3) in the way cities and infrastructure are planned, built, organized and governed; 4) in the way our economies and industries function and 5) in the way our societies function on a local, national and international level.

The earlier these changes are implemented, with an emphasis on equity and justice, the more options and solutions for Climate Re-

silient Development will become available. We don't have any time to lose. As our report has also shown, missing the Paris Agreement goals will limit our options for a sustainable and climate resilient future, because a warmer world brings higher climate impact risks to which people and places will have to adapt. We know now that securing a healthy and climate resilient future for all is difficult, if not impossible to achieve in many regions, with global warming higher than 2°C over the medium and long term.

FAQ 6: What is Climate Resilient Development and how do we pursue it?

Worldwide action to achieve a climate resilient, sustainable world is more urgent than previously thought. But what can be done? Our report highlights a solutions framework that we call Climate Resilient Development. It combines strategies to adapt to climate change with actions to reduce greenhouse gas emissions to support sustainable development for everyone. Action to implement this concept has to start now because making progress is already challenging at current global warming levels. If temperatures exceed 2°C of warming, climate resilient development will become impossible in some regions of the world.

We know that climate change presents risk to nature, people and infrastructure around the world. These risks will increase with every small increase in warming, and reducing them is made more complicated by other global trends such as over-consumption, population growth, rapid urbanization, land degradation, biodiversity loss, poverty and inequity, etc. In short: the world is facing a long list of complex and interacting challenges that need to be dealt with simultaneously.

Both the urgency and the complexity of the climate change crisis require actions at a new depth and scale. Our report provides a solutions framework that successfully combines strategies to deal with climate risks (adaptation) with actions to reduce greenhouse gas emissions (mitigation) which result in improvements for natures and people's well-being – for example by reducing poverty and hunger, improving health and livelihoods, providing more people with clean energy and water and safeguarding ecosystems on land, in lakes and rivers and in the ocean. This solutions framework is called Climate Resilient Development.

In developing countries and in areas that are particularly exposed to climate change (e.g. in coastal areas, small islands, deserts, mountains and polar regions) climate impacts and risks can exacerbate vulnerability and injustices which can undermine efforts to achieve sustainable development, particularly for marginalized communities. But each country will follow its own path. Importantly, each country has different capacities and opportunities for Climate Resilient Development. Nevertheless, our report clearly shows that rapidly scaled-up, well-aligned investment facilitates Climate Resilient Development and that it advances more quickly with increased international cooperation and financial assistance.

Striving for Climate Resilient Development means reducing exposure and vulnerability to climate hazards, cutting back greenhouse gas emissions and conserving biodiversity are given the highest priorities in everyday decision-making and policies on all aspects of society including energy, industry, health, water, food, urban development, housing and transport. It is about successfully navigating the complex interactions between these different systems so that action in one area does not have adverse effects elsewhere and opportunities are harnessed to accelerate progress towards a safer, fairer world.

Climate Resilient Development isn't achieved with a single decision or action. It's the result of all of the choices we make about climate risk reduction, emissions reductions and sustainable development on a daily basis. New scientific evidence shows that addressing the risks and impacts of climate change successfully involves a more diverse set of actors than previously thought – it not just policy-makers but everyone in government, civil society and the private sector. For example, if we consider changes in agriculture, it takes a combination of effective government policy and regulation as well as informed daily decisions by farmers, traders and agricultural companies to lead to fundamental change which is required to adapt to a changing climate, reduce greenhouse gas emissions and secure lives and livelihoods not just of those directly involved but for wider society as well.

In brief: Climate Resilient Development involves everyone. The prospects for effective action improve when governments at all levels work with citizens, civil society, educational bodies and scientific institutions, the media, investors and businesses and form partnerships with traditionally marginalised groups, including women, youth, Indigenous Peoples, local communities and ethnic minorities. In such a societal setting, scientific, Indigenous and local knowledge and practical knowhow can come together to provide more relevant effective actions. In addition, different interests, values and worldviews can be reconciled if everyone works together.

Targeting a climate resilient, sustainable world involves fundamental changes to how society functions, including changes to underlying values, worldviews, ideologies, social structures, po-

litical and economic systems, and power relationships. This may feel overwhelming at first, but the world is changing anyway and will continue to change so Climate Resilient Development offers us ways to drive change to improve well-being for all – by reducing climate risk, tackling the many inequities and injustices experienced today, and rebuilding our relationship with nature.

The choices we make in the next decade will determine our future. Our report clearly states Climate Resilient Development is already challenging at a warming level of less than 1.5°C, and will become more limited by 2°C. In some regions, it will be impossible if the temperature exceeds 2°C, including low-lying coastal cities, settlements and small islands, some mountain areas and polar regions.

This key finding underlines the urgency for climate action and that focusing on equity and justice as well as on adequate funding, political commitment and partnerships lead to more effective climate change adaptation and emissions reductions.

The scientific evidence is unequivocal: climate change is a threat to human well-being and the health of the planet. Any further delay in concerted global action will miss a brief and rapidly closing window to secure a liveable future.

Appendix IV
Causes of Water Pollution

Other Water Pollution Causes

(Environmental Pollution Centers (2022).

The **causes of water pollution** vary and may be both natural and anthropogenic. However, the most common causes of water pollution are the anthropogenic ones, including:

Agriculture runoff - carrying fertilizers, pesticides/insecticides/ herbicides and other pollutants into water bodies such as lakes, rivers, ponds). The usual effect of this type of pollution consists of algae growing in affected water bodies. This is a sign of increased nitrates and phosphates in water that could be harmful to human health.

Storm water runoff – carrying various oils, petroleum products, and other contaminants from urban and rural areas (ditches). These usually forms sheens on the water surface.

Leaking sewer lines – may add trihalomethanes (such as chloroform) as well as other contaminants into groundwater ending up contaminating surface water, too. Discharges of chlorinated solvents from dry-cleaners to sewer lines are also a recognized source of water pollution with these persistent and harmful solvents.

Mining activities – mining activities involve crushing rocks that usually contain many trace metals and sulfides. The leftover material from mining activities may easily generate sulfuric acid in the presence of precipitation water.

Foundries – have direct emissions of metals (including Hg, Pb, Mn, Fe, Cr and other metals) and other particulate matter into the air. Please, read more at Foundry.

Industrial discharges – discharges produced by industrial sites may add significant pollution to water bodies, but are usually regulated today.

Accidental leaks and spills – associated with handling and storage of chemicals. They may happen anytime and, although they are usually contained soon after they occur, the risk of polluting surface and groundwater still exists. An example is ship accidents such as the Exxon Valdez disaster, which spilled large amounts of petroleum products into the ocean.

Deliberate/illegal discharges of waste – while such occurrences are less common today, they may still happen due to the high cost of proper waste disposal; illegal waste discharges into water bodies were recorded all over the world.

Burning of fossil fuels – the emitted ash particles usually contain toxic metals (such as As or Pb). Burning will also add a series of oxides including carbon dioxide to air and, respectively, water bodies.

Transportation – even though Pb has been banned in gasoline in the U.S. and many other countries, vehicle emissions pollute the air with various tailpipe compounds (including sulfur and nitrogen compounds, as well as carbon oxides) that may end up in water bodies via deposition with precipitation water.

Construction activities – construction work can release a number of contaminants into the ground that may eventually end up in groundwater.

Plastic materials/waste in contact with water – may degrade slowly releasing harmful compounds for both human health and ecosystem.

Disposal of personal care products and household chemicals (including detergents and various cleaning solutions) – this is a serious problem since these releases into water are unpredictable and hard, if not impossible to control. It is up to each of us to minimize this contribution to water pollution by controlling our consumption and disposal of such chemical products, as well as trying to recycle as much as we can!

Improper disposal of car batteries and other batteries – may add metals

Leaking landfills – may pollute the groundwater below the landfill with a large variety of contaminants (whatever is stored by the landfill).

Animal waste – contribute to the biological pollution of water streams.

About the Author

Author: Dr. Ronald Barnes, PhD (2023)

Ronald Barnes received a B.A in Economics, from Coe College in Cedar Rapids, Iowa. Dr. Barnes is a member of the Coe College Athletic Hall of fame.

He received his Masters' degree in Religious Studies from the University of Chicago and studied 1 year at Arizona State University in Political Psychology.

Mr. Barnes is completing his PhD in Psychology (2023).

Mr. Barnes spent over 15 years in corporate America prior to establishing his own business. After 10 years in business Mr. Barnes returned to school for a M.A and PhD.

Mr. Barnes has authored three previous books entitled. "Practice what you preach, preach what you practice" and "the Pyrrhic State of America: Make America Great", and "An American Reality". Mr. Barnes is an inducted member of Psi Chi, the international psychology honor society, and a member of Kappa Alpha Psi fraternity.

Every human should have a clear understanding of environmental issues confronting our world. Climate change / global warming is an issue that is forefront to the everyday phenomena humans confront. The nature of the Universe is to respond to the conditions (stimulus) with which it is confronted. Over the years humans paid little to no attention to the affect progress has on the world. Now we are aware of the fact that progress, as humans have pursued it, comes with a price. The issue is not to discontinue progress, but to now progress in ways that are compatible with the sustenance and continuity of our world.

www.ingramcontent.com/pod-product-compliance
Lightning Source LLC
Chambersburg PA
CBHW071546210326
41597CB00019B/3137

9781957114712